MEN

against

The Problem of Battle Command in Future War

by S. L. A. MARSHALL

COLONEL, AUS

Historian of the European Theater of Operations

GLOUCESTER, MASS.

PETER SMITH

1978

REPRINTED, 1978, BY PERMISSION OF
Mrs. S. L. A. MARSHALL

PRINTED IN THE UNITED STATES OF AMERICA

ISBN 0-8446-4057-3

CONTENTS

To my old friends and comrades of the Historical Division,
War Department Special Staff,
and of the Historical Division,
European Theater of Operations, because they combined
a strong sense of duty with an eager
and imaginative questing for knowledge.

AUTHOR'S NOTE - 1961

AT FORT BENNING, GA., the citadel of the infantry spirit in modern America, there is today a superior system for bringing infantry weapons under command control so that in the crisis of battle their response will be decisive. It is called *Train Fire*.

As a system, scientifically applied over the known distance ranges, so that riflemen and integral support weapons groups will acquire fire habits which will fortify them and keep unity of action steadfast when they face the enemy, Train Fire owes its origins formally to the Human Resources Research Office of the Army and to the imaginative tacticians who advise that body of scientists.

But its progenitors, in line tail-mane as a cavalryman might say, are this book, that part of the U.S. infantry which together faced fire during the Korean War, 1950-53, and a little-known but greatly loved soldier of this country, Colonel Wayne Archer.

Archer should have been known as Mr. Infantry to the Army of World War II. He was an officer with broad vision, large sympathy and rare command capacity, though fortune being what it is, he never attained high rank. He originated the Infantry Combat Badge and inspired many other reforms designed to lift the prestige of the rifle line, though the Army has all but forgotten these things. In a very real sense, "Men Against Fire," is Colonel Archer's book, and now that it is appearing in new dress, my greatest satisfaction is that at long last I may acknowledge my debt to him.

In the European Theater, 1944-45, he was chief of the Senior Observers Board. Most of the members were elderly colonels. They were supposed to determine by direct witness how our tactics and fire system worked and recommend change where needed. Most of them were too old for that; they were willing enough to face danger, but they had lost the knack of cultivating the confidence of young America, which is not overly-communicative at best, and is least of all so when it undergoes trial by fire. My people, on the other hand, were also observers, of the same age group as the men of a rifle squad, and they had to work in the fire zone whether they liked it or not. *C'est la guerre.*

So very early in the ETO, Archer and I arrived at an *ad hoc* arrangement which had no blessing from higher authority. My people would develop the body of data, covering the organization of river crossings, the effectiveness of river crossings, methods of patrolling, etc., etc. This would be handed to Archer and his people would study it and arrive at the judgments for transmission to Ground Force, the War Department, et al.

We worked it that way until V-E Day and it resulted in one of the most stimulating friendships of my lifetime. Inevitably, it threw Archer and me into as frequent and intimate association as our own tours of the front permitted. The maximum of my own time which administrative affairs permitted was spent forward with the infantry formations because I had discovered during my 10 months in Central Pacific operations that many of our assumptions about the human equation and group response under fire were radically wrong. To learn more about wherein and why we miscalculated seemed the most profitable way to spend my time.

The data which came of this prolonged personal research was my own and I made no attempt to cross-check or co-relate it with the findings of my friends and colleagues in the Historical Division, ETO. There was a reason for this quite apart from the lack of time and the high pressure of duty. Each man judges performance according to some standard deriving from his own experience. But the impressions of others, and how they evaluate man against fire, are also either validated by a breadth of experience or colored by a lack of perspective. Where the armchair historian may pick and choose whatever fits into the making of a good story, the combat historian may be sure only of his own datum plane.

What I learned more or less on my own, I talked over at great length only with Wayne Archer, except for one evening when I discussed the factor of combat isolation with the Supreme Commander, General Dwight D. Eisenhower. Invariably, Archer listened carefully, asking hardly a question, and at the end he would say: "Marshall, this is all new ground. We have suspected some of these things about infantry. But we have never known for sure. You have it pinned down. Now you must write it. For if you don't, it may never be said."

Our talks were always relaxing. The praise usually terminated an evening of drinking together. Whether the words made me feel better than the scotch is beyond saying. At first the latter was taken more seriously. Then there came an evening when Archer (the time was March, 1944) was extra mellow and extra serious. Just before we said good night he insisted that we solemnly shake hands while I promised out loud that I would begin writing the book on the day I started my terminal leave. He said: "Unless you have a plan, you won't do it." Well, we got that over with, but when I awakened bright-eyed next morning, the promise was only dimly registered.

Separation came in May, 1956. I was at the Pentagon, anticipating a holiday in Florida. Archer was still in Europe. On that day there came to my office a messenger with a special delivery letter from him, the gist of which was: "You promised me. Now do it!" There was no choice but to comply and the writing of the book was how I celebrated my deliverance from things military.

I had anticipated that, provided the book had something important to say, the American people might gradually be won to its truths, and over 10 years or so, their good opinion would superinduce change in the Army toward needed reform. It was a naive expectation. There was practically no public response to the book. Civilian periodicals ignored it almost unanimously. But within less than six months, the United States Army, and other military systems abroad, had taken it up seriously, and such fortune as it has had, they made.

True, in the beginning, the most noteworthy reaction was that some of the older generals (especially those who had been division commanders) read it but to damn it. They felt compelled to believe that whatever it said about troops performing somewhat less than perfectly under fire, might hold true of some other commander's people, but never of their own. Of such loyalty as this great legends are made, but it is hardly conducive to the smoothing of operations and the correcting of error. The grumbling hurt no one and after the grumblers had heard from some of their own battalion and company commanders, who being released from service, were ready to sound off immoderately about these same problems, they quickly lapsed into silence, as did certain of the service schools which initially objected that the criticisms were too radical.

The Army institutionally immediately welcomed the book and put the main ideas to use within the training system. With especial gratitude, I recall the support given it by General J. Lawton Collins, Chief of Staff, General Jacob Devers, chief of field forces, and Lieutenant General Raymond S. McLain, then Army chief of information. The department publication, "Officers' Call," gave it a first ringing endorsement.

Out of the text, what was said about fire ratios—that less than 25 percent of our infantry line employed hand weapons effectively when under fire—drew main attention and stirred initial controversy. The data so said and the Army didn't contest it. Instead, at centers like Forts Benning, Knox and Riley, during the years 1948-49, to overcome weapons inertia, imaginative trainers instituted wholly new methods, some of which were suggested in the book.

Their pioneering paid off splendidly in Korea from 1950 till the end. When on returning from my first tour there, in April, 1951, I reported to Secretary of the Army Frank Pace, Jr., and General Charles L. Bolte, vice chief of staff, I was able to say that active weapons participation in our infantry line had risen beyond 55 percent both in night defense and daytime attack—more than doubling the World War II output.

General Bolte then raised the question: "Is this not because the perimeters are isolated and the men know they must shoot to survive?" To that I perforce answered: "No, knowledge of the conditions would but increase the impact of fear, which freezes the trigger finger. The improvement has to be the product of an improved training system." What was most noticeable in Korea was that every infantry company was aware of the problem. Their fire volume was a point of pride with them. In World War II, junior leaders, without exception, disregarded the factor.

As a system, *Train Fire* marks still another advance. Its techniques are described in Army pamphlet No. 355-14, which is circulated for the information of troops. On page 1 are to be found these words: "Less than 25 percent of men in combat (or 1 out of four) fired their weapons at all." So in 1958 was re-confirmed officially the figure with which I first startled Wayne Archer in 1944. But we had to know it before we could start upgrading.

This leads us to the final point. One of the simplest truths of life is that it is possible for a problem of major dimension to exist within fighting bodies (or any other organization) and remain unrecognized for years until one person points it out. It must be circumscribed before there is belief in its existence. We often preach about the virtue of completed staff work; but we seldom tell the junior executive that this means command of his data above all else. It should encourage him to know also that the jungle is full of diamonds awaiting the digger and that there is more to be learned about men against fire than has ever appeared between the covers of books.

S. L. A. MARSHALL
BG, USAR, Rtd.

Dherran Dhoun
Birmingham, Michigan

AUTHOR'S NOTE - 1947

THESE commentaries touch mainly on certain physical and psychological aspects of the problem of command in minor tactics. They treat of the future of war only in so far as it is necessary to establish that these aspects should be the objects of earnest research. In writing them I have been guided by two general considerations:

(1) I have tried to avoid repeating most of the basic and valid ideas which are to be found in the many excellent books on leadership and on the training of soldiers for war. (2) I have tried to dismiss, or to disprove, any of the clichés and practices pertaining to the preparation of fighting men which, it seems to me, have nothing in their favor except that they are traditional.

One of the deterrents to the adoption of new concepts is that company officers and non-coms rarely write of their combat experiences. Even when they do so they are unlikely to search into the reason and nature of them, usually because their experiences are narrow and personal. Also, they have no way of gauging what things are typical and characteristic.

In consequence, most of our textbooks and commentaries on leadership and the mastery of the moral problem in battle are written by senior officers who are either wholly lacking in combat experience or have been for long periods so far removed from the reality of small arms action that they have come to forget what were once their most vital convictions and impressions. It is very easy to accept dogma and to dilate upon it and dress it in modern uniform when one views battle from an armchair in a training headquarters or when one reaches the comfortable state of believing that because certain ideas have worked in the past, it is not necessary to ask whether they can be improved upon.

Ever since World War I, it has seemed to me that we have been pulling in opposite directions in many of our basic policies governing the preparation of our man power for combat, and the several reasons for most of the confusion in the military mind, as well as in the public mind, on this matter all stemmed from one source.

By and large, our training system and our standard of battle discipline still adhere to the modes of the eighteenth century, though we are working with the weapons and profess to be working with the advanced military ideas of the twentieth. That this is so will be made clear, I trust, in the course of this book. It is not intended in a spirit of criticism. The Army of the United States assigned me the best opportunity ever given any soldier to observe how the masses of our men react in battle and to measure the common denominators of our weakness and our strength in close combat. It was inevitable that these fresh data would point to new conclusions, some mildly unorthodox. I have written mainly in the hope that I will be able to give something back and that the ideas found here will jog the minds of thinking soldiers and earnest civilians alike. While I have dealt primarily with the problem of ground forces, I have related it to the subject of our future national security. To this extent, the book has an especial claim on the interest of the non-military public to whom its subject matter is being presented for the first time.

My indebtedness to the thousands of men and officers of the Army of the United States whose experiences, and whose candor, made possible the success of the field work on which this book is based is indeed beyond statement. Some few of them I have mentioned by name because it seemed consistent with the purposes of the text. There are hundreds of others who were equally helpful. Numbers of them are now dead. Those losses will not be vain if the lessons which come of their battlefield experience are heeded toward a perfecting of the arms of the United States.

Finally, I believe it a duty to say that in the course of my work, no combat commander from the regular establishment ever put a stone in my path. In the overwhelming majority, they were willing and eager to assist in the search for battlefield truth, even when the facts hurt. No one group was more helpful than the younger battalion and regimental commanders, and staff officers, only a few years out of United States Military Academy. I cannot help but reflect on these things when I read the criticisms of captious individuals who are ever ready to condemn the West Point system on one frivolous ground or another instead of inquiring into its ability to build steadfast character into young manhood, which is the main issue.

S. L. A. Marshall
Detroit, Michigan

1

THE ILLUSION OF POWER

"Our business, like any other, is to be learned by constant practice and experience; and our experience is to be had in war, not at reviews."

—SIR JOHN MOORE.

IN THE early years of World War II, it was the common practice of public spokesmen in the United States to magnify the role of the machine in war while minimizing the importance of large forces of well-trained foot soldiers.

When France fell this became a favorite theme of the American press and radio and the few voices which were raised in protest against it were scarcely heard. The idea took hold in the Congress. Makers of military policy argued that we should concentrate on air power and armored force, almost to the exclusion of infantry. Ultimately common sense prevailed and we struck a tolerably good balance. But so strong was the influence of the machine upon our thinking, both inside and out of the military establishment, that as the new Army took shape, the infantry became relatively the most slighted of all branches. In the assignment of man power to other arms and services we allowed a sufficient margin. We did not do so with infantry.

The effects were almost catastrophic. It forced us in the

European Theater of Operations to become the first army in modern history to undertake a continuing and decisive operation without the shadow of an infantry reserve. That the Supreme Commander and his Staff accepted and mastered this risk must be rated as among the highest of their achievements. Even so, the strain which the situation put upon command is a small matter when compared to the almost incredible powers of endurance which it exacted of the divisions in line. Once committed, there was no choice but to keep them in action. The only relief afforded them was when they were moved from one part of the combat zone to another.

Some of the facts in the record are almost unbelievable. On August 6, 1944, exactly two months after the invasion of Normandy, the entire re-enforcement pool for infantry forces in Europe—here I speak of infantrymen ashore in France and ready to go into battle—consisted of one lone rifleman. I repeat, there was only one man at hand to replace the many who were being killed on that day. That was the low point so far as the immediate reserve in France was concerned. But the situation with respect to infantry replacements arriving from or training in the United States grew steadily worse. By November the Theaters were scraping the bottom of the barrel. Battle was continuing to take its daily toll from the infantry divisions then in line. As in World War I, some of our hastily trained re-enforcements were arriving at the front without a working knowledge of their weapons. But while the strain on the front grew worse by the hour, in the rear there were no more infantrymen, either at hand or in sight or promised.

How real was this crisis? It was so very real that in the middle of the Ardennes fighting in late December, 1944, the governing condition that made certain of the commanders, including Marshal Sir Bernard Montgomery, hesitate and argue for postponement of the counteroffensive was the non-availability of American riflemen re-enforcements. That was the sum and substance of Montgomery's general misgiving about the condition and situation of the First United States Army, which had recently come under his Army Group command. The story that Montgomery overrated the power of the German attack, that his blood turned to water, and that he saw no alternative but to withdraw to behind the line of the River Meuse is a piece of headquarters gossip which has no foundation in fact. He objected only on the ground which has been here stated. The validity of his objection was soon supported by the turn of events, for in the middle of the battle, and under the threat of defeat, our authorities reached their belated decision to succor the infantry situation by training infantry re-enforcements from among the surplus man power which had been assigned to the Air and Anti-Air services. In this way the crisis was passed.

The propaganda which had sought the practical elimination of foot forces as a major factor in mobile war was thoroughly injurious, however well intended. Throughout the war, it reacted as a depressant upon the self-esteem of our infantry forces, thereby reducing their general combat efficiency. This fact is established by Army polls which show that our infantrymen, in the great majority, continued to hold a low opinion of the importance of their own branch.

However, the really curious thing about the prophets who were so ready to proclaim the supreme importance of the machine and the relative unimportance of trained man power in modern warfare was that they had remained singularly blind to the most obvious conditions of the war which was already in progress. Had they but studied these conditions more closely, they would have observed that the effect of machine war was already to increase the masses of mobilized man power beyond anything previously known in history. Not only were there more millions of men under arms. Within the armies then engaging there were larger musterings of infantry forces than ever before.

The nature of the weapons of modern warfare made such a balance inevitable. The mode of warfare and the make-up of armies are never determined by arbitrary choice. The conditions of war are fixed primarily by the civilization of the period. The character of a society, the inventive genius of its people, and the productive potential of its lands and cities determine finally the choice of weapons in war. In turn, it is the choice of weapons which regulates whether armies shall be large or small and whether national defense can be delegated to a compact and highly mobile professional army or must be entrusted to the national mass, under professional guidance.

These are the dominating and unalterable considerations. In a well-roaded and therefore accessible industrialized civilization, where such weapons as the medium tank and heavy bomber made it certain that the flanks and rear of a national society would come under constant and severe attack, there was never the slightest possibility that the issues between great nations could be settled by limited forces in a thunderclap of action along their frontiers. General Charles de Gaulle dreamed of such a thing, but he did not dream very clearly. Marshal Hans von Seeckt likewise advocated it in his theory of Lightning War. But the context of his writings establishes that he believed in it only as the precursor to the total war between national societies.

When World War II began, it was self-evident that the only possible form of the conflict in its decisive stages would be total war. The illusions of the Sitzkrieg period did not deceive those who kept their minds on weapons rather than on politics, realizing that in war the latter is shaped by the former. The embattling of an entire society was certain to necessitate defense by the entire society, to the limit of its material and man-power resources.

But it is unfortunately the case that the masses of men are not capable of taking other than a superficial judgment on the effect of new weapons. History records, moreover, that their military leaders do not always see and think clearly in such matters. As great a soldier as U. S. Grant was slow to understand the revolutionizing effect of the rifle bullet upon tactics. For more than a generation following the Civil War, our naval experts could foresee development of the armored vessel only in the form of a ram. The failure of higher commanders in World War I to understand the potential of armored power and to make proper tactical application of it is an example of almost incredible blundering.

Yesterday's lesson underscores the moral for today. Once the total contest between national societies is predicated, it becomes impossible to write off the ultimate clash between the masses of men who fight on foot. They are the body of the national defense. If foresight has not already assured their prompt and efficient mobilization, the emergency will compel it. In the hour when decision is made possible through the attainment of a superiority in the striking (fire) power of the heavy weapons of war, they must go forward to claim the victory and beat down the surviving elements of resistance. There is no other way out. The society which looks for an easier way is building its hope on sand.

The belief in push-button war is fundamentally a fallacy. But it is not a new fallacy. It is simply an age-old fallacy in modern dress. There is one controlling truth from all past wars which applies with equal weight to any war of tomorrow. No nation on earth possesses such limitless resources that it can maintain itself in a state of perfect readiness to engage in war immediately and decisively and win a total victory soon after the outbreak without destroying its own economy, pauperizing its own people, and promoting interior disorder.

War must always start with imperfect instruments. Equally, these instruments can never be fully perfected in the course of war. There are fixed limits to the national resources and within these limits each element of the national defense competes with all others. Too, once the sides are drawn, each side must reshape and balance its own mechanism according to what appears to be the point of greatest vulnerability in the other. The Germans in the last war were markedly short of field and siege artillery and of motorized supply. But they had expected to win before there was need of these things. The Japanese were short in many materials, among them such basic needs as wire and explosives. We were short not only of infantry but of motorization. The tightest checkrein of all upon our general operations was the shortage of landing craft. And so it will be in the future. Improvisation is the natural order of warfare. The perfect formulas will continue to be found only on charts.

Though these broad propositions may appear to be so clear and simple as to be generally acceptable, the difficulty arises in the attempt to apply them to any new situation. Today we have such a situation. The atom bomb has conditioned and clouded all thinking about general military problems. Fear is uppermost. The public imagination, caught by the flights of fancy of self-appointed spokesmen who have taken snap judgment on the problem of the new weapon, is ready to believe that atomic power has negated all military truths. The fatal idea continues to spread that nothing counts except the future use, or non-use, of this one weapon.

Yet the principle remains inviolate: *The existence of any weapon which fully jeopardizes a whole society necessitates the readiness for defense by the whole society.*

One reads the fatuous prediction by a national columnist that the next war will be won by five men smuggling atom bombs in suitcases through a customs line. Another writer of national repute makes the equally preposterous comment that the wars of the future will be won by scientists and engineers and will have little use for soldiers.

The fact remains that no society can afford to risk its fate on the possibility that decision may be obtained quickly by use of the new weapons, though their power is such that they reduce cities to rubble in the twinkling of an eye. Least of all can our society afford such a risk. Our tradition and our law remain as we have known them. We have not forsworn our national character because of the bomb. We will not wage aggressive war upon our neighbors. Therefore we can never count on striking the first blow.

The fatal flaw in the logic of any belief in quick victory is that it is but one side of the coin, the other side being the submission to quick and final defeat in the event that the first round goes badly. If we are not willing to accept this alternative risk, then the vaporings of those whose main appeal is to the fears and imaginings of the multitude rather than to the national intellect cannot be permitted to prevail against the immutable common-sense truths of war which have been confirmed by the experience of man throughout the ages.

So long as men live on this earth, and until the arrival of the hour when there is a weapon of such lethal capacity as to make it evident beyond doubt that the whole race of man can be blotted out by the turning of a switch, it must be reckoned that the prime effect of the increase in the killing power of weapons is to increase the need for man power in the national defense.

But it is not solely with respect to this over-all aspect of modern war and the effect of its weapons that we are called upon closely to re-examine the position. Just as in industry the machine has brought fresh and untold confusion to the problems of human relationship, in armies it has transformed the problem of the human equation without at the same time provoking its essential re-estimate. I do not refer to social relationships within an army, which is another though not less important facet of the general problem, but to the tactical relationships of men in battle.

Since more than a century ago, when the rifle bullet began its reign over the battlefield and soldiers slowly became aware that the day of close-order formations in combat was forever gone, all military thinkers have pondered the need of a new discipline. It has been generally realized that fashioning the machine to man's use in battle was but half of the problem. The other half was conditioning man to the machine. The mechanisms of the new warfare do not set their own efficiency rate in battle. They are ever at the mercy of training methods which will stimulate the soldier to express his intelligence and spirit.

The philosophy of discipline has adjusted to changing conditions. As more and more impact has gone into the hitting power of weapons, necessitating ever widening deployments in the forces of battle, the quality of the initiative in the individual has become the most praised of the military virtues. It has been readily seen that the prevailing tactical conditions increased the problem of unit coherence in combat. The only offset for this difficulty was to train for a higher degree of individual courage, comprehension of situation, and self-starting character in the soldier.

From this realization have come new concepts of discipline which have altered nearly all of the major aspects of life and of human association within western armies. We have continued to grapple with the problem of how to free the mind of man, how to enlarge his appreciation of his personal worth as a unit in battle, how to stimulate him to express his individual power within limits which are for the good of all. It is universally recognized that as the means of war change, so must the intelligence of man be quickened to keep pace with the changes.

Our weakness lies in this—that we have never got down to an exact definition of what we are seeking. Failing that, we fall short in our attempt to formulate in training how best to obtain it, and our philosophy of discipline falters at the vital point of its practical, tactical application.

I say that it is a simple thing.

What we need in battle is more and better fire.

What we need to seek in training are any and all means by which we can increase the ratio of effective fire when we have to go to war.

The discipline, the training methods, and the personnel policies of our forces should all be regulated to conform with this one fundamental need. In whatever we do to mold the thought of the combat soldier, no other consideration should be given priority ahead of this decisive problem.

Today we are once again at the parting of the ways. We have come through another great war and its reality is already cloaked in the mists of peace. In the course of that war we learned anew that man is supreme, that it is the soldier who fights who wins battles, that fighting means using a weapon, and that it is the heart of man which controls this use. That lesson we are already at the point of forgetting. We can ill afford it.

Since my return to the United States in January, 1946, I have been astonished at the number of my civilian friends who have told me pointedly that it is folly to write about the experience of World War II except in the measure needed faithfully to record the facts of the national history for the benefit of those who have purely an academic interest. They are already certain that the outline of the next war is shaping up as something entirely different. They doubt that any of the tactical and human lessons of the past will continue to apply. They believe that it is ox-cart thinking to dwell now on the importance of the human element in close combat.

With these conclusions I disagree. Further, I believe that they are so completely wrong that they constitute a positive danger to the future security of the United States.

The last war and the launching of the atom bomb may have shattered an epoch. But no man yet knows what will emerge from its ruins. The choice is not wholly in our hands. As threatening as is the shadow which this new form of military power casts across humanity, it is not more terrible than the prospect that we will misread its meaning and risk the national future on a theory of warfare which cannot be sustained in the event that we are challenged. We have seen proof of the great killing power of the new weapon. We know that it can shatter a sitting fleet. Such is our awe that there is room for little else in the landscape. This is not the course of wisdom. It is still necessary to study and to perfect a plan of general operation around our major weapons. But we begin with a false estimate of their value if we already discount those other vast problems of national security which by their nature remain relatively static.

The favorite quotation of Britain's great military critic, Captain Liddell Hart, has it that "we learn from history only that we do not learn from history." Once again, this remains to be proved.

As a military historian I believe that the lessons of the last war are as pertinent to our study of how to prepare in the future as were the lessons of World War I to our situation before 1939. That means including the lessons of Hiroshima and of the rocket attacks on Great Britain, making due allowance for the rapid evolution of these forms of offensive warfare. But it does not mean concentrating on these things to the exclusion of many other vital truths from our own experience.

After World War I, we examined the evidence and we saw mainly that the multiplying of automatic weapons had given a preponderanec of power to the defensive. We neglected to project the evolutionary progress of air and armored power and we failed to estimate how these forces would compel a new balancing of the equation. The danger now is that we will reverse this error and make a mistake of the opposite kind.

I see no profit in introducing any argument as to the probabilities of our present civilization achieving a real disarmament, followed by a long peace, and finally, the outlawing of war. That is outside my sphere. Suffice it to say that from what I have read of history and observed of man's nature, I deem this to be such a slender chance that the national fate cannot be risked upon it. Yet no man truly serves his country today who will not work to bring this hope nearer.

This book deals with future war only so far as is necessary to establish that the battlefield remains a living reality. There is always the chance that the reader, though he may agree that the end of war is unlikely, may be one of those who believe that hand fighting is as dead as the cavalry arm. I hold this to be a fallacy and in the next chapter I have explained why. The ideas advanced there are not invalidated in the event that we succeed in outlawing the atomic bomb and the guided missile weapons. The shadow of these weapons will remain over mankind. What has been done before can be done again. Future war will always move toward totality, once nations engage. Either that, or we are nearing the millennial dawn.

With this exception, I shall confine the discussion largely to the tactical fact which is at once the simplest and the most complex topic in the military art—man himself as a figure on the field of combat. It is my belief that he has been too long neglected.

2

ON FUTURE WAR

"In 1919 I was the sole person who saw war in the form it would be; yet saw it only as an acorn and not as an oak."

—MAJOR GENERAL J. F. C. FULLER.

THE battlefield is the epitome of war. All else in war, when war is perfectly conducted, exists but to serve the forces of the battlefield and to assure final success on the field.

It is on the battlefield that the issues of war are decided. Yet it may well occur in the struggle between nations that such a preponderance of power will be achieved by one side or the other, or such destruction will be worked on one body or the other either by the weapons of the air or by naval blockade, as to virtually predetermine the results of the battlefield.

Even so, the contest between land armies will continue to be the concluding act in war. Without this conclusion, military victory will not be achieved.

The greater becomes the emphasis on weapons whose destructive power is aimed at the civil society, the more certain it is that the battlefield will continue as the area of final decision in war. This trend cannot be reversed. It can be ended only when the mortal dangers to all civil populations are so universally recognized, and that recognition is so directly reflected in the policies of the various states and the attitudes of their peoples, as to end war itself.

The mobilizing of all national forces and resources in war does not lessen the decisive importance of all that occurs on the battlefield. Nor can the evolution of new weapons establish a form of war in which military decision is foreseeable and the danger of stalemate can be reckoned avertable, without full preparation to engage the land forces of the enemy with forces of the same sort.

The over-all effect of the changing pattern of war, as it is superinduced by the character of the new weapons, is to promote an ever increasing involvement of national forces and national prestige. This in turn makes more critical the events of the battlefield. For it should be noted that all military power is dependent on the civil will. It is the nation and not its army which makes war. But when all of the forces of a society are directed toward the shaping of a decisive instrument in war and the cutting edge of the instrument fails on the field of battle, the result is not alone the defeat of an army, but the envelopment or dissolution of a society.

The objective in an international war between great states will continue to be the complete subjugation of one or more nations. "Unconditional Surrender" was not a condition unique to the times of Casablanca or to the ideological character of the war which was being waged upon us by the Axis nations. It was the natural derivative of what is called "total war," this being a state of war in which all of the assets and aspects of the lives of nations are faced with attack, and in consequence, all elements and resources of the engaging peoples must be subject to use in the defense.

Whether wars between nations become total in character

or are conducted on a limited scale by moderately sized military forces is not determined finally by the rapacity of either side, though it is a common illusion that such is the case. The condition is fixed by the range and the hitting power of the dominant weapons, coupled with the ability of the state and people to sustain this power.

The possession of weapons which make possible direct attack upon the very heart of society during the early stages of conflict is the factor which assures the totality of wars between great states. As witness, until June, 1940, the world in general believed that the issue between France and Britain on one side and Germany on the other might be decided in a limited war. The warring nations so believed: there was no intense speed-up of production on either side. During the first nine months, the military, political, and social forces within the two democracies reacted sluggishly to the emergency. The period was conspicuous for its lack of direct military attack against the defending societies. But when the mass and range of the German hitting power and the purpose of the German state to give them maximum use were made clear by the overrunning of France and the air blitz on Great Britain, both sides intensified their preparations for total war. The USSR and the United States, which had not entered the war, did likewise.

The character of the new weapons with which the United States is armed—weapons such as the atom bomb, the B-36 bomber, the guided missiles and the implements of bacteriological warfare, all of which have the civil society as target—predispose that should we engage another major power, the war will be total. The trend of all weapon development today is toward increasing the attack upon cities; in the next war far greater damage will be done to population centers, with infinitely less risk to the attackers. This is the only thing of which we may be reasonably certain, since the fullness of the aim in total war also has the effect of militating against the prospect for quick victory.

Our possession of, and predilection with, weapons which aim at the civil population as the best means of immobilizing the enemy forces, in turn predisposes that any nation which seeks to engage us will be prepared for total war. This is the natural political consequence of the evolution of armament and the consequence cannot be averted short of the outlawing of long-range weapons. As this appears altogether unlikely, we should accept the fact that the armament advance of other major air powers will roughly parallel our own. Every improvement in weapon power is aimed at lessening the danger on one side by increasing it on the other. Consequently, every improvement in weapons is eventually met by a counter-improvement. In an age of warfare which has as its chief characteristic unlimited air or guided missile attack upon relatively defenseless targets, a chief effect of this competition in armament is that while reducing the chance for quick victory by either side, it compounds the deadliness of war for both sides.

We have already seen this proved. Prior to World War II, the exponents of mechanized power and of air power both contended that their weapons would make war less costly in the long run, not only in lives but in money. Theirs was a simple line of reasoning: Swift, sure movement made certain swift, sure victory. The course of the war showed that they were wholly mistaken. Their weapons performed with even greater tactical efficiency than they had expected, but their error lay in the fact that they had misunderstood the nature of war and misread the history of its development. All that they achieved was to set the stage for total war. Far from altering this fundamental, the development of atomic power assures its continuation into the future.

The true objective not only of the atomic weapon but of

rockets and modern bombing fleets is the physical destruction of a society, just as in limited war the true objective of short-range weapons was the annihilation of its military forces. This will continue to be not only the most profitable and vulnerable target but the actual object in war, and it cannot be changed by humanitarian declarations of policy or by international agreement. To suggest that these super weapons should be aimed at military installations only would be like bringing up the heavy artillery to shoot at a clay pipe; they are designed, primarily, for no such limited target.

The cities are a profitable target, first, because they produce for war, second, because they are the transport and supply bottlenecks of a national system, and third, because more people can be killed there; they are a vulnerable target because they must remain exposed. The masses of people, the factories, and the communications of a society cannot go underground. On the other hand, the counter-hitting military weapons, along with their personnel and maintenance establishments may well do so. Their reserves may be either protected in the same way or put in the remoter sections of the interior.

We see here, already in process, a curious transposition whereby the civil mass becomes the shield covering the body of the military, and wherein the prospect for final military success lies in the chance that the shield will be able to sustain the shock, and sufficient of the will and productiveness of the civil population can be maintained until the military body can make decisive use of its weapons. It is bootless to protest that civilization will not tolerate such a danger and that therefore atomic power will be brought under control, since it is all too clear that the determining conditions are above and beyond the existence of the atomic weapon.

Yet even with atomic power, and making due allowance for further development of this and other weapons, it must be deemed impossible to work physical destruction upon an entire society. Space defeats such a possibility, even if the bomb were not subject to the same limitation known to every other lethal weapon, namely, that the precision if its use can never be made equal to its destructive power.

It must follow, however, that the only target short of the objective (physical destruction of the society) which will satisfy the aim of the new air weapons is the moral and spiritual obliteration of the society. The will of the mass must be wholly subjugated through the physical damage worked on the body of the society and through the destruction of all prospect for success by the military forces. For until this end is reached, military resistance will be continued and victory will be denied.

To say that full surrender of the mass will is requisite in future wars between great states is only another way of stating that unconditional surrender will be a normal requirement for the peace. The victor will determine at his own peace table whether he wills the survival of the vanquished state.

What then is unconditional surrender by a state? It is the surrender of every last bargaining right by the people and their representatives. We saw this happen in the case of Germany. That it did not happen with either Italy or Japan was because it suited the political purposes of the victors to will otherwise.

However, with the totality of the state endangered and unconditional surrender of the society being the object in war, all power within the state will be mobilized for war immediately after its outbreak. Once the emergency arises, the society will demand nothing less, even though it be a society which now imagines that the new weapons provide a short cut to victory and which wishfully believes, or fatalistically fears, that the future has no place for the masses which fight on foot. All of the available man power within the state, meaning all men and women of possible military use beyond those required for war production, civil defense, and the maintenance of the interior economy, will be formed into armies and into other forces for the common defense.

If it be correct that the nature of weapons prefixes the conditions of total war and unconditional surrender, there can be no alternative to the mobilization of maximum forces. They must stand ready to protect the frontier if the air weapons of the enemy gain early ascendancy and his forces prepare to invade, and they must be equally ready to invade, seize, and occupy ground if one's own weapons have made the enemy vulnerable.

In fact, since full advantage must be taken of the dislocation and destruction worked by these weapons, all ground forces must be ready to exploit their use by moving into enemy country with much greater celerity than ever before. The only logical strategic corollary of decisive strength in the air arm is the movement by air of all forces which fight on ground—infantry, tanks and artillery—and of the supply necessary to sustain them. This logic breaks down only at the point where the question arises whether it is economically possible to develop and maintain an air transport of such capacity. That it would be decisively advantageous is incontestable.

For it should be well noted that out of total attack and defense must come total conquest of the enemy ground and total occupation of his lands and cities. What we now see in Germany and Japan is the pattern for the aftermath of any great war of the future.

There is only one other possible outcome and it is not less fearsome—that such a war may be fought to a stalemate in which both sides are defeated because of mutual destruction of the means which would permit of military decision. To consider future war as a contest between great air powers, whose strength in the overhead attack has not been balanced with proportionate strength in ground tactical forces, is to open the door to this final absurdity. It makes the whole idea of air war as banal as a suicide pact.

Thus if we are to attain to such balance in our planning for the national defense as to assure that our military undertakings can proceed toward their proper end, which is political action, it must be reckoned that the battlefield continues as one of the realities of war.

By their nature the new weapons make all areas, however remote from the frontiers, a possible scene of waste and of death. Our newest bomber can travel 10,000 miles with a pay load; the range of rockets is expected to reach 7000 miles in the not distant future. One effect of these developments is to make the battlefield relatively less dangerous while the menace of war to the life of ports, communications centers, and industrial cities rises steadily.

But it is not the fact of death and of killing which is the prime characteristic of the battlefield. Its essential is that it is the meeting place of opposing military forces where they engage in decisive struggle for the possession of ground.

The forces of the battlefield possess the means of attack and of defense and the balancing of these two forms of warfare is their whole preoccupation. It is when these forces move to within killing range of each other with the flat trajectory weapons, and when they put these weapons to use for the purpose of killing, that the battlefield becomes defined. There is no battlefield until two forces close, each with the object of overriding the body of the enemy while avoiding being overridden.

It is my belief that the field, as I have defined it here, has not lost its decisive character, and that in the nature of things it cannot do so.

It may well be that the battlefield will not be the chief killing ground in future war. It is conceivable, also, that the nature of the preponderant weapons may so change the shape of wars to come that decision will appear almost as an event of anticlimax.

But decision in war is a clinching act. It is the action which finally delivers the victory surely into one's hands. That which is decisive cannot be measured simply in terms of how the preponderance of force is weighted within the victorious side. Nor does it come simply of counting the opposing rows of the dead.

Decision implies a final determination of the issue. It is obtained by those who survive and not by those who die in striving for it. It is an act which brings about a final submission by the enemy and the restoration of political action. It is an advance on Richmond, not a Gettysburg, a bold stroke across the bridge at Remagen, not a landing on the coast of Normandy.

In total war, decision recedes further and further into the distance until one final act brings about quick collapse and submission of the force protecting the enemy interior. Be the chaos of the defending civilization ever so great, as long as there remains an organized will to resist, defeat is not insured. The final act will always be an act of the battlefield, whether the ground forces which achieve it move by overland transport or by sea or by air.

Air power is essential to national survival. But air power unsupported by the forces of the battlefield is a military means without an end.

3

MAN ON THE BATTLEFIELD

"On the battlefield the real enemy is fear and not the bayonet or bullet. All means of union of power demand union of knowledge."

—ROBERT JACKSON.

IT IS of the battlefield, as I have defined it earlier, that I speak in saying that the mind of the infantry soldier should be conditioned to an understanding of its reality through all stages of his training.

He needs to be taught the nature of that field as it is in war and as he may experience it some day. For if he does not acquire a soldier's view of the field, his image of it will be formed from the reading of novels or the romance written by war correspondents, or from viewing the battlefield as it is imagined to be by Hollywood. One of the purposes of training should be to remove these false ideas of battle from his mind.

To give the soldier a correct concept of battle is a far different thing from encouraging him to think about war. The latter is too vast a canvas; it includes too much detail which is confusing to his mind and immaterial to his personal problem.

We have surpassed all other armies and outstripped common sense in our effort to teach the man something about war. He is counseled about war's causes, which is a good thing on those rare occasions when the instruction is in qualified hands. He is told about how the soldiers and sailors of other nations observe courtesy and foster tradition. He is even bored by lectures on the strategy and logistics of high command.

But he does not get what he most requires—the simple details of common human experience on the field of battle. As a result, he goes to the supremely testing experience of his lifetime almost as a total stranger.

Those facts which are denied him should be made his not only for the sake of personal survival but in the interests of unit efficiency. The price for failure is paid all up and down the line; men go into action the first time haltingly and gropingly, as if they were lost at night in the deep woods. Lives are wasted unnecessarily. Time is lost. Ground that might be taken is overlooked.

It is not necessary that these misfortunes befall organizations simply because they are new to battle.

It is possible that the infantry soldier can be trained to anticipate fully the true conditions of the battlefield; it is possible that units can be schooled to take full and prompt action against the disunifying effect of these conditions. Fear is ever present, but it is uncontrolled fear that is the enemy of successful operation, and the control of fear depends upon the extent to which all dangers and distractions may be correctly anticipated and therefore understood.

I feel sure that a majority of my professional readers will agree that these things are so, but will protest that the protective measures have always been taken. There are certain of the facts of life, we have long said to one another, which can only be learned the hard way. Let me therefore anticipate the direction in which these protests will be leading.

It is true that the individual soldier in the more recent periods of warfare has been trained to regulate his movements on the field of battle according to the nature of the ground. He has been schooled to maneuver with his weapons in such a way that his employment of the ground will give his weapons maximum effectiveness and himself a degree of protection. That is the desirable physical equation for each man going into combat with the purpose of firing against the enemy—to find an efficient site for the weapon which is at the same time a relatively secure site for the firer.

"Surely," it will be said, "this is the heart of the matter—the relating of the weapon to the ground and of the soldier to the weapon and the ground, and the relating of all weapons within the formation to each other and to the ground, so that there will be maximum fire power and maximum defensive strength within the position."

My answer to this fundamental proposition in traditional military logic is that it is absolutely false.

The heart of the matter is to relate the man to his fellow soldier as he will find him on the field of combat, to condition him to human nature as he will learn to depend on it when the ground offers him no comfort and weapons fail. Only when the human, rather than the material, aspects of operation are put uppermost can tactical bodies be conditioned to make the most of their potential unity.

In the course of this book, I propose to show in detail wherein our tactics are unnecessarily weakened because we do not consider human nature in our fire training, and I propose to show also that the required adjustments are as definable as the adjustments of a machine gun or any other mechanism.

It is beyond question that the most serious and repeated breakdowns on the field of combat are caused by failure of the controls over human nature. In minor tactics the almost invariable cause of local defeat is fundamentally the shrinkage of fire. In the greater number of instances this shrinkage is the result of men failing to carry out tasks which are well within their power. That is what I mean by failure of control. But it is to be noted that the responsibility for this failure is shared by all alike. It does not imply a weakness simply in command. The additional control which is needed is that kind which is requisite when any one or two or more individuals must undertake a difficult and dangerous task together and it is necessary that they proceed with an economy of effort. Toward that end it is essential that the will of one give direction to the mission even though there be not more than two in the working unit.

It is my belief that a system of man-to-man control on the battlefield is our great need in tactics and that it is fully attainable. This is not a metaphysical problem. It can be attacked by rather simple methods, once the factors of the problem are understood. We all grant that the soldier must be trained for initiative and encouraged to think about his personal problem while in combat. Too, we are at the opening of a new age in warfare when it appears certain that all operation will be accelerated greatly, that all ground formations must have greater dispersion for their own protection, and that therefore thought must be transmitted more swiftly and surely than ever. These things being true, it is an anachronism to place the emphasis in training and command primarily on weapons and ground rather than on the nature of man.

A careful study of past military history and particularly of the "little picture" of our own infantry operations in the past war leads to the conclusion that weapons when correctly handled in battle seldom fail to gain victory. There is no other touchstone to tactical success, and it is a highly proper doctrine which seeks to ingrain in the infantry soldier a confidence that superior use of superior weapons is his surest protection. However, in modern infantry warfare the correct use of weapons by a formation in battle comes of the perfecting of controls over men who are physically beyond reach. While we all recognize this in principle, it is my belief that we have not applied its lessons sufficiently to our training system and that we are still under the spell of ancient training doctrines even though we disclaim their objectives.

Our training methods are conditioned by the ideal of automatic response. At the same time, our observation of the battlefield's reality makes clear to us that we need men who can think through their situation and steel themselves for action according to the situation. Under the conditions of national service, there is not time to instill in the infantry soldier that kind of discipline which would have him move and fire as if by habit; but even if there were time for such training, it would be unsuited for an age of warfare which throws him upon his own responsibility immediately combat starts.

There are two roads open and they lead in opposite directions. Our difficulty is that we try to move both ways at one time. The thinking soldier—the man who is trained for self-starting—cannot be matured in a school which holds to the vestiges of the belief that automatic action is the ideal thing in the soldier. Discipline, long and assiduously applied, may inculcate such a degree of automatic response in a soldiery that the majority will do as told when ordered. Or training may endeavor to teach all men to think clearly and to school them in methods which will strengthen mutual resolve so that when an emergency occurs, the majority will think and act correctly, though no general order is given. But these are mutually exclusive ends of a training system; it is the responsibility of training to make a clean choice and then hew to the line. Far from encouraging a retreat from the basic disciplinary idea of "I command: you obey," I am suggesting that it is chiefly when command is exercised as if it were based on some military magic known only to officers that it precludes that form of obedience which is distinguished by intelligent and aggressive action.

The soldier can be conditioned to make full use in combat of his fellow man. This psychological objective is by no means beyond the possibility of attainment, if the problem is approached simply and with courage.

During training, the soldier, and certainly the officer, can be given enough knowledge about human nature under the stresses of the battlefield that when it comes his time to go forward, he can make tactical use of what he knows in the same way that he applies what he has learned about his equipment.

To point him toward this new field of understanding would not make him less amenable to instruction in the employment of weapons and the use of ground. Rather, every line would be underscored. He would begin to understand the importance of weapons in the light of their combat limitations as well as their uses, and fresh warmth would be added to matters which he now regards as coldly mechanical.

What I advocate, in brief, is the substitution of reality for romance in all discussion of the battlefield, and the introduction into training in the maximum measure possible of the same element which steadies a command during its trial by fire.

On the field of fire it is the touch of human nature which gives men courage and enables them to make proper use of their weapons. One file, patting another on the back, may turn a mouse into a lion; an unexpected GI can of chocolate, brought forward in a decisive moment, may rally a stricken battalion. By the same token, it is the loss of this touch which freezes men and impairs all action. Deprive it of this vitalizing spark and no man would go forward against the enemy.

I hold it to be one of the simplest truths of war that the thing which enables an infantry soldier to keep going with his weapons is the near presence or the presumed presence of a comrade. The warmth which derives from human companionship is as essential to his employment of the arms with which he fights as is the finger with which he pulls a trigger or the eye with which he aligns his sights. The other man may be almost beyond hailing or seeing distance, but he must be there somewhere within a man's consciousness or the onset of demoralization is almost immediate and very quickly the mind begins to despair or turns to thoughts of escape. In this condition he is no longer a fighting individual, and though he holds to his weapon, it is little better than a club.

It has happened too frequently in our Army that a line company was careless about the manner in which it received a new replacement. The stranger was not introduced to his superiors nor was there time for him to feel the friendly interest of his immediate associates before he was ordered forward with the attack. The result was the man's total failure in battle and his return to the rear as a mental case.

So it is far more than a question of the soldier's need of physical support from other men. He must have at least some feeling of spiritual unity with them if he is to do an efficient job of moving and fighting. Should he lack this feeling for any reason, whether it be because he is congenitally a social misfit or because he has lost physical contact or because he has been denied the chance to establish himself with them, he will become a castaway in the middle of a battle and as incapable of effective offensive action as if he were stranded somewhere without weapons.

This is a basic principle in the elementary psychology of the infantry soldier. Though I have personally investigated several hundred of the heroic exploits by single individuals in the past war and have read many stories of the hair-raising feats of long-handed guerrillas, I have yet to find the episode which is at odds with it.

In the fiction of warfare, C. S. Forester's *Rifleman Dodd* is a splendid and almost monolithic figure. But let us read the story again. The initiative of Dodd did not flow from the unhelped spirit of one man. Finally, he was as dependent upon his Spanish companions as were they upon him.

It is that way with any fighting man. He is sustained by his fellows primarily and by his weapons secondarily. Having to make a choice in the face of the enemy, he would rather be unarmed and with comrades around him than altogether alone, though possessing the most perfect of quick-firing weapons.

4

COMBAT ISOLATION

"The finest theories and most minute plans often crumble. Complex systems fall by the wayside. Parade ground formations disappear. Our splendidly trained leaders vanish. The good men which we had at the beginning are gone. Then raw truth is before us."

—MAJOR GENERAL CHARLES W. O'DANIEL.

SOMEWHERE along in the course of this writing it is necessary to describe the battlefield as it must seem to a stranger. Else, much that is said here will not make sense.

The battlefield is cold. It is the lonesomest place which men may share together.

To the infantry soldier new to combat, its most unnerving characteristic is not that it invites him to a death he does not seek. To the extent necessary, a normal man may steel himself against the chance of death.

The harshest thing about the field is that it is empty. No people stir about. There are little or no signs of action. Over all there is a great quiet which seems more ominous than the occasional tempest of fire.

It is the emptiness which chills a man's blood and makes the apple harden in his throat. It is the emptiness which grips him as with a paralysis. The small dangers which he had faced in his earlier life had always paid their dividend of excitement. Now there is great danger, but there is no excitement with it.

That is what makes the going tough. A man builds himself up to the realization that danger will come on him suddenly. He thinks a lot about how he will react to the shock of knowing that he is under fire.

In his mind's eye he imagines a situation with himself as center. He will be afraid but he will be stimulated. It will be like participating in a tough team game. While it lasts he will at least get some warmth from it and he will be supported by the strength that he feels all around him.

But it doesn't work out that way. Instead, he finds himself suddenly almost alone in his hour of greatest danger. And he can feel the danger, but there is nothing ou there, nothing to contend against. It is from the mixture of mysification and fear that there comes the feeling of helplessness which in turn produces greater fear.

That is what green troops are up against. Time and again I have heard them say after their first try at combat: "By God, there was never a situation like it. We saw no one. We were fighting phantoms." And as frequently they have added this, though their convictions about the matter were wholly at odds with the facts of situation: "We had to do it all alone. We got no support on either flank."

Follow the course of an average infantryman through training to his first grapple with the enemy and it is easy to see why these illusions persist and why they plague the effort to rally moral strength in men their first time in battle.

In training the soldier grows accustomed to the presence of great numbers of men and of massive mechanical strength close around him. He sees this strength on parade. He watches it on maneuvers, and though this last is supposed to be the closest approximation of war, he never feels lonely in the field. The more he sees of the strength of the army, the greater grows his confidence, though he is scarcely aware that it has become a factor in his morale.

Even the forces of the enemy are virtually materialized for him. He watches their formations maneuver in the training movies. Pictures of their squads adorn the day room walls. From captured films, the sequences of which have been arranged so as to convey the idea of mass and of strength, he gathers an impression of how they look when they engage in battle. This is the target! This is what he will meet some day! They are rugged, but they are flesh-and-blood, fully mortal, and therefore vulnerable. There is nothing mysterious about it. If the target is hit by a bullet, it will go down.

There are other impressions but these are the main ones. He believes what he has seen and his belief hardens his imaginings of what his own first experience will be like.

He thinks of battle as the shock impact of large and seeable forces, a kind of head-on collision between visible lines of men and machines extending as far as the eye can see. It is quite vain to say that he should have known better and that he should have realized that the need of guarding while hitting sharply limits the spectacle of engagement. The fact is that he does not know and that all the vicarious impressions of battle which he has gained since childhood fix the other picture in his mind.

During the advance of his company toward the zone of fire, nothing happens to modify his original impressions. Troops in great number and matériel in almost inexhaustible quantity are among the common aspects of the rear area. One feels the power of an army even more strongly there than elsewhere.

Somewhere along on the road up, this soldier hears the distant sounds of battle, but they are the impersonal sounds of heavy guns so they are only mildly disturbing to his emotions. They produce no dispersion in the force right around him and he is given no cause to reflect on the dwindling of the appearance of strength close at hand. True, he can see now mainly the strength which is in his own column and he has no idea what friendly forces may be moving up on right and left along parallel routes. That his own outfit is grouped around him is enough; every man close at hand is an aid in helping him choke down the fear which might otherwise have stopped him.

The unit enters upon the battlefield and moves across ground within range of the enemy's small arms weapons. The enemy fires. The transition of that moment is wholly abnormal. He had expected to see action. He sees nothing. There is nothing to be seen. The fire comes out of nowhere. He knows that it is fire because the sounds are unmistakable. But that is all that he knows for certain.

He is less sure now of what he had expected would steady him in just such a crisis—the fire power and unity of his own outfit. The men scatter as the fire breaks around. When they go to ground, most of them are lost to sight of each other. Those who can still be seen are for the most part strangely silent. They are shocked by the mystery of their situation. Here is surprise of a kind which no one had taught them to guard against. The design of the enemy has little to do with it; it is the nature of battle which catches them unaware. Where are the targets? How does one engage an enemy who does not seem to be present? How long will it be until the forces opposite begin to expose themselves and one's own forces will rally around the tactical ideas which training had taught them would prove useful? How long until engagement begins to assume its normal aspect?

There is none present to tell this rifleman or his comrades that this *is* normal and that only his personal reaction to it may change with time. He may go on and on through repeated engagements and never know a situation that is more tangible. In essence, it is against this very situation that his unit must find the means to rally if it is to succeed in battle. There may come days when the field is alive with action and visible targets are plentiful and the supporting strength of one's own side is plainly visible on flanks and rear. But these are the characteristics of movement and of breakthrough. They will become salient only after fire has won the decision or has assured the winning of it.

The enemy fire builds up. Its aim becomes truer. The men spread farther from each other, moving individually to whatever cover is nearest or affords the best protection. A few of them fire their pieces. At first they do so almost timidly, as if fearing a rebuke for wasting ammunition when they do not see the enemy. Others do nothing. Some fail to act mainly because they are puzzled what to do and their leaders do not tell them; others are wholly unnerved and can neither think nor move in sensible relation to the situation. Such response as the men make to the enemy fire tends mainly to produce greater separation in the elements of the company, thereby intensifying the feeling of isolation and insecurity in its individuals. The junior leaders are affected as much as the rifle files. The unexpectedness of the experience has made them less confident, and the more confidence slips, the more they hesitate to give orders which might stimulate action by the more aggressive men. That orders are not given furthers the demoralization and immobility of the line. Knowing that the leaders are afraid makes the men more fearful.

Could one clear commanding voice be raised—even though it be the voice of an individual without titular authority—they would obey, or at least the stronger characters would do so and the weaker would begin to take heart because something is being done.

But clear, commanding voices are all too rare on the field of battle. So they wait, doing nothing, and inaction takes further toll of their resolve. More grievous losses will no doubt come to this band of men in time, but as a company this is the worst hour that they will ever know.

Their losses will become their great teacher. The weaker ones will be shaken out of the company by this first numbing experience, adding fresh numbers to the statistics which show that more battle fatigue cases come from initial engagements than from all subsequent experience in the line. Some who might have been saved, had great wisdom been given those who were responsible for their training, will go to this scrap heap. A majority of the strong will survive. In the next round with the enemy they will begin to accustom themselves to the nature of the field and they will learn by trial and error those things which need doing to make the most of their united strength.

It would serve no purpose to dwell on the discouraging detail of this ordeal if it were not for the belief that much of it is unnecessary and that the infantry soldier can find a better way.

One must not come to rest on Clausewitz's gloomy warning that: "In war the novice is only met by pitch black night." On beyond that are to be read the words: "It is of first importance that the soldier, high or low, should not have to encounter in war things which, seen for the first time, set him in terror or perplexity."

That is the desired goal—to shed such a strong light in training that it will dispel much of the darkness of battle's night. We have the word of the nineteenth century's great military thinker that it can be done. It remains a hope for those of us who weigh the military problem of the new age.

5

RATIO OF FIRE

"All this is but a sheep in a lion's mouth except the breed and disposition of the people be stout and warlike."

—Francis Bacon.

NOW I do not think I have seen it stated in the military manuals of this age, or in any of the writings meant for the instruction of those who lead troops, that a commander of infantry will be well advised to believe that when he engages the enemy not more than one quarter of his men will ever strike a real blow unless they are compelled by almost overpowering circumstance or unless all junior leaders constantly "ride herd" on troops with the specific mission of increasing their fire.

The 25 per cent estimate stands even for well-trained and campaign-seasoned troops. I mean that 75 per cent will not fire or will not persist in firing against the enemy and his works. These men may face the danger but they will not fight.

But as I said in the beginning, it is an aspect of infantry combat which goes unheeded. So far as the records show, the question has never been raised by anyone: "During engagement, what ratio of fire can be expected from a normal body of well-trained infantry under average conditions of combat?"

This is a very curious oversight, inasmuch as the problem of how much fire can be brought to bear is the basic problem in all tactics. In fact, it *is* tactics in a nutshell, and the other elements of tactics are simply shaped around it.

Commanders in all ages have dealt with this central problem according to the weapons of their day and their imaginative employment of formations which would bring the maximum strength of these weapons to bear at the decisive point. Surely that is the heart of the matter so far as the mechanics of battle are concerned—to arrange men, to move them, to counter-move them, so that their own ranks will have a lesser exposure while their weapons are exploiting a greater vulnerability in the ranks of the enemy. Great Cyrus of Persia was thinking on these things when his scouts warned him that the Egyptian phalanx was one hundred men deep. For he answered them: "If they are too deep to reach their enemies with their weapons, what good are they?"

Yet almost without exception such doubts as were expressed by the great minds which have pondered tactics and shaped their development through the ages have touched mainly upon the geometry of the problem. They were all optimists, these distinguished captains, and they appear to have taken it for granted that if they could devise a superior pattern and plan of maneuver, the willing response of well-trained troops would correspond very closely to the number of spear points which could get at the body of the enemy or the number of muskets which were in position to fire.

We see an occasional slight doubt expressed on the subject. There was the time when Marshal Maurice de Saxe computed the rate of fire in light infantry. "To cross three hundred yards will take six to seven minutes. Light infantry should be able to fire six shots a minute, but under the stress of battle four should be allowed for. Each man will consequently fire thirty shots during the advance. As there are seventy men, two thousand shots will altogether be fired."

There, certainly, is a famous estimate to conjure with—a 66 per cent efficiency in the rate of fire in the early days when the musket was an extremely crude weapon and armies were largely composed of mercenaries. De Saxe was far off base. He would not get that rate of fire today doing the same movement with seventy soldiers armed with the M-1 and carbine. On the range, certainly, but advancing against an enemy who was defending with small arms fire, never!

Surely it is a curious oversight that the questions of how much fire can be brought to bear and how fire ratios may be increased have been treated throughout history as if the solutions were to be found only in terms of mechanics and geometry. One who reads these tactical doctrines would be justified in concluding that it is a point of honor with professional soldiers to hold dogmatically to the belief that training conquers all, and that when perfectly drilled and disciplined, all men will fight.

But what of human nature? In the workshop or office, or elsewhere in the society, a minority of men and women carry the load of work and accept the risks and responsibilities which attach to progress; the majority in any group seek lives of minimum risk and expenditure of effort, plagued by doubts of themselves and by fears for their personal security.

When the deeper currents of life run counter to the proposition that a majority of men will engage willingly, it would not appear reasonable to believe that military training will succeed where other disciplines fail. "The gravest problem of the commander," Lieutenant General Robert C. Richardson has commented, "is to train his younger subordinates to close the circuit. All may be loyal, but not one in four is sufficiently self-disciplined to see each task through to its proper finish."

If this be true of the officer corps under the normal pressure of garrison responsibility, what is to be expected of the ranks in the hour of danger when the "closing of the circuit" entails additional self-exposure in order to return the fire of the enemy?

Why the subject of fire ratios under combat conditions has not been long and searchingly explored, I don't know, but I doubt that it is because of any professional taboo, and I suspect that it is because in earlier wars there had never existed the opportunity for systematic collection of the data.

It is the human nature of the commander to believe that the majority of his troops are willing, for unless he so believes, he is aware of his personal failure. But it is not less true that to his mind willingness and loyalty are virtually synonymous with initiative and voluntary risk at the point of danger. During battle it is physically impossible for him to make a check of the action of all of his men without neglecting other and more decisive responsibilities. Nor can his immediate subordinates do this for him without taking undue risks. After battle the question is of less moment and the commander becomes occupied with the duties of his next employment. It does not occur to him that the rate of effective fire in the command is the core of his whole problem and that the means for taking a reasonably accurate measure of it is his for the asking. The average soldier will tell the absolute truth when asked if he has used his weapon.

In the course of holding post-combat mass interviews with approximately four hundred infantry companies in the Central Pacific and European Theaters, I did not find one battalion, company, or platoon commander who had made the slightest effort to determine how many of his men had actually engaged the enemy with a weapon. But there were many who, on being asked the preliminary question, made the automatic reply: "I believe that every man used a weapon at one time or another." Some added that wherever they had moved and viewed, it had seemed that all hands were taking an active part in the fighting.

Later when the companies were interviewed at a full assembly and the men spoke as witnesses in the presence of the commander and their junior leaders, we found that on an average not more than 15 per cent of the men had actually fired at the enemy positions or personnel with rifles, carbines, grenades, bazookas, BARs, or machine guns during the course of an entire engagement. Even allowing for the dead and wounded, and assuming that in their numbers there would be the same proportion of active firers as among the living, the figure did not rise above 20 to 25 per cent of the total for any action. The best showing that could be made by the most spirited and aggressive companies was that one man in four had made at least some use of his fire power.

Naturally, the commanders were astonished at these findings, though at the conclusion of the critiques, there was no case of a commander remaining unconvinced that the men had made a true report.

Most of the actions had taken place under conditions of ground and maneuver where it would have been possible for at least 80 per cent of the men to fire, and where nearly all hands, at one time or another, were operating within satisfactory firing distance of enemy works. Scarcely one of the actions had been a casual affair. The greater number had been decisive local actions in which the operations of a company had had critical effect upon the fortunes of some larger body and in which the company itself had been hard-pressed. In most cases the company had achieved a substantial success. In some cases, it had been driven back and locally defeated by enemy fire.

The critiques covered all that took place from the opening to the end of action. The spot checks were made by a showing of hands and questioning as to the number of rounds used, targets fired upon, etc., usually after all witnesses had been heard and the company had received a well-rounded impression of the action as a whole. There is no reason to doubt that the men were reporting honestly and objectively; they quickly realized that it was something to their credit if they could establish that they had participated in the fire fight.

There was an occasional exception to the almost uniform pattern of the results but there was no exception to my earlier statement that the commanders had not been trained to interest themselves in this problem.

To return to the beginning, in the Makin Island fight, which was a part of the Gilbert Islands invasion in November, 1943, one battalion of the 165th Infantry Regiment was stoutly engaged all along the front of its defensive perimeter throughout the third night. The enemy, crazed with sake, began a series of banzai charges at dusk, and the pressure thereafter was almost unremitting until dawn came. The frontal gun positions were all directly assaulted with sword and bayonet. Most of the killing took place at less than a ten-yard interval. Half of the American guns were knocked out and approximately half of the occupants of the forward foxholes were either killed or wounded. Every position was ringed with enemy dead.

When morning brought the assurance that the defensive position had weathered the storm and the enemy had been beaten back by superior fire, it seemed certain to those of us who were close enough to it to appraise the action that all concerned must have acted with utmost boldness. For it was clear that the whole battalion was alive to the danger and that despite its greatly superior numbers, it had succeeded by none too wide a margin. We began the investigation to determine how many of our men had fought with their weapons. It was an exhaustive search, man by man and gun crew by gun crew, each man being asked exactly what he had done.

Yet making allowances for the dead, we could identify only 36 men as having fired at the enemy with all weapons. The majority were heavy weapons men. The really active firers were usually in small groups working together. There were some men in the positions directly under attack who did not fire at all or attempt to use a weapon even when the position was being overrun. The majority of the active firers used several weapons; if the machine gun went out, they picked up a rifle; when they ran out of rifle ammunition, they used grenades. But there were other witnesses who testified that they had seen clear targets and still did not fire.

It is true that these were green troops who were having their first taste of combat. Likewise, it is to be observed that the nature of perimeter defense, as it was then used in the Pacific, limited the freedom of fire of troops inside the perimeter.

But thereafter the trail of this same question was followed through many companies with varying degrees of battle experience, in the Pacific and in Europe. The proportions varied little from situation to situation. In an average experienced infantry company in an average stern day's action, the number engaging with any and all weapons was approximately 15 per cent of total strength. In the most aggressive infantry companies, under the most intense local pressure, the figure rarely rose above 25 per cent of total strength from the opening to the close of action.

Now maybe I should clarify the matter still further. I do not mean to say that throughout an engagement, the average company maintained fire with an average of 15 per cent of its weapons. If that were it, there would be no problem, for such a rate of fire would necessarily mean great volume during the height of an assault.

The thing is simply this, that out of an average one hundred men along the line of fire during the period of an encounter, only fifteen men on the average would take any part with the weapons. This was true whether the action was spread over a day, or two days, or three. The prolonging of the engagement did not add appreciably to the numbers.

Moreover, the man did not have to maintain fire to be counted among the active firers. If he had so much as fired a rifle once or twice, though not aiming it at anything in particular, or lobbed a grenade roughly in the direction of the enemy, he was scored on the positive side. Usually the men with heavier weapons, such as the BAR, flamethrower or bazooka, gave a pretty good account of themselves, which of course is just another way of saying that the majority of men who were present and armed but would not fight were riflemen.

Terrain, the tactical situation, and even the nature of the enemy and the accuracy of his fire appeared to have almost no bearing on the ratio of active firers to non-firers. Nor did the element of battle experience through three or four campaigns produce any such radical change as might be expected. The results appeared to indicate that the ceiling was fixed by some constant which was inherent in the nature of troops or perhaps in our failure to understand that nature sufficiently to apply the proper correctives.

One of the principal effects of battle seasoning is apparently to make junior leaders cognizant of some of the proportions of the problem so that when the company engages, a larger percentage of NCOs will use direct methods to increase the fire power of the immediate group. But the best of NCOs cannot for long move up and down a fire line booting his men until they use their weapons. Not only is that an invitation to sudden death but it diverts him from supporting and encouraging the relatively few willing spirits who are sustaining the action. Also, regardless of what the book says to the contrary, that is not his real role on the battlefield. When the heat is on, he is more likely to be working hard with his own weapon, trying to beat back the enemy with his own hands and strength of purpose.

It seems to me, therefore, that there is every reason why the fire ratio factor should be treated primarily as a most vital training problem and secondarily as a subject for critical inquiry and treatment in the early stages of combat.

During the Kwajalein battle, in working with the companies of the 7th Infantry Division, we first found that the percentage of men who engaged with all weapons was about constant in all companies, despite extreme variations in the local tactical situations. Then attention was drawn to one other significant fact. Though there were a few minor shifts, with new men coming forward and others leaving the fight because of death or wounds, in the main the same men were carrying the fire fight for each company day after day. The willing riflemen, grenadiers, and bazooka men who had led the attack and worked the detail of destruction upon the enemy on a Monday would carry the attack when the fight was renewed in a different part of the island on Wednesday. The hand that pulled the trigger was the same hand that was most likely to be found tossing a grenade, setting a satchel charge, or leading a sortie in the next round.

Of course there were many other active files doing yeoman service in supply, communications, and other missions. Men do not progress in battle by fire alone, and without the work of the others the efforts of the firers would have been unavailing. But the point is that among those present for duty with the weapons, the same names continued to reappear as having taken the initiative and relatively few fresh names were added to the list on any day.

You could pick out your man who would probably keep going until he was dead. Or for that matter, after a few trial rounds, you could spot the man who would probably never get going though his chances of dying were relatively good.

For it must be said in favor of some who did not use their weapons that they did not shirk the final risk of battle. They were not malingerers. They did not hold back from the danger point. They were there to be killed if the enemy fire searched and found them. For certain tasks they were good soldiers. Nor can it be doubted that as riflemen many of them were of sound potential. The point is that they would not fire though they were in situations where firing was their prime responsibility and where nothing else could be as helpful to the company.

It was also conspicuous that the men who used their weapons were the same men who took the lead in outflanking an enemy fire trench or in blowing an enemy shelter. It should be obvious that these things go hand-in-hand, since the act of willingly firing upon the enemy is of itself an instance of high initiative on the battlefield, though commanders have long considered it as simply a natural derivative of sound training. On that subject I will have more to say later.

How much then does training have to do with it? Probably this—that it enables the willing soldier, the man who will fight when he gets the chance, to recognize the breadth of each opportunity and to know when and where to use his fire to full advantage and with regard for his own need of protection. It may also stimulate and inform the man who is already fixed with a high sense of duty so that in him the initiative becomes simply a form of obedience.

But more than that it is not likely to do under present methods and until the principles by which we attempt to establish fire discipline are squared with human nature. We are on infirm ground when we hold to the belief that the routine of marksmanship training and of giving the soldier an easy familiarity with his weapon will automatically prompt the desire to use the weapon when he comes under fire.

There is no feature of training known to any company commander I have met which enabled him to determine, prior to combat, which of his men would carry the fight for him and which would simply go along for the ride. Discipline is not the key. Perfection in drill is not the key. The most perfectly drilled and disciplined soldier I saw in World War I was a sergeant who tried to crawl into the bushes his first time over the top. Some of the most gallant single-handed fighters I encountered in World War II had spent most of their time in the guardhouse. It is all very well for such an authority as Major General J. F. C. Fuller to assure us that the yardsticks of loyalty and obedience are the means of measuring beforehand the probable response of the soldier in battle. Many others have said it before Fuller. But I deny that it is true. It may have applied to the ranks in the days of closed formations but it does not apply to our present soldiery.

We had better face the facts of life. Fire wins wars, and it wins the skirmishes of which war is composed. Toss the willing firers out of an action and there can be no victory. Yet company by company we found in our work that there were men who had been consistently bad actors in the training period, marked by the faults of laziness, unruliness, and disorderliness, who just as consistently became lions on the battlefield, with all of the virtues of sustained aggressiveness, warm obedience, and thoughtfully planned action. When the battle was over and time came to coast, they almost invariably relapsed again. They could fight like hell but they couldn't soldier.

Did these earlier signs of indiscipline then provide any light in the search for men who would probably act well in battle? Not at all! Fighting alongside the rough characters and taking an equally heroic part in the actions were an even greater number of men whose preliminary conduct had marked them as good soldiers. In the heat of battle these forceful individuals gravitated toward each other. The battle was the pay-off.

Almost unexceptionably in the company actions which we brought under survey in the course of combat historical work, there were a number of private soldiers whose earlier service had been lusterless, but who became pivots of strength to the entire line when fire and movement were needed, exhibiting all of the enterprise and judgment of good junior leaders. Numerous witnesses attested how the sustained action of these men had rallied others around them. We found, however, that the company commanders did not always know how these men had served them, and that they did not as a rule understand that these signs of exceptional battlefield strength should be studied as factors in a re-evaluation of how best to employ the company strength in battle. Yet it cannot be gainsaid that the whole moral strength of a fighting command pivots around the men who are willing to employ their weapons against the enemy, and if these men were not present, the company would be morally a cipher.

During the mass interview of one infantry company in the Central Pacific, the statements of all concerned made it evident that one of the sergeants had performed so conspicuously during two days of attack that the progress of the company had pivoted largely around his individual exploits. He had not been recommended for decoration though the facts revealed by the others said clearly that he was deserving of the

Silver Star three times over. I asked the company commander why nothing had been done. He frowned his astonishment at the question.

"When the fighting started, he practically took the company away from me," he answered. "He was leading and the men were obeying him. You can't decorate a man who'll do that to you."

So I asked him if he had ever reckoned with the fact that no commander is capable of the actual leading of an entire company in combat, that the spread of strength and the great variety of the commander's problems are together beyond any one man's compass, and that therefore a part of his problem in combat is to determine which are the moral leaders among his men when under fire, and having found them, give all support and encouragement to their effort. He said no, that no one had ever told him these things.

And though these ideas are basic in command, that young captain was not alone in his narrow outlook. There were many others who seemed to regard command as a prerogative rather than as a responsibility to be shared with all who are capable of filling up the spaces in orders and of carrying out that which is not openly expressed though it may be understood. Even where a proper initiative had been exercised (as in the sergeant's case, since it was evident that his actions had always bettered the situation of the company) they were prone to question any action by a junior which might be construed as an encroachment on their authority.

These were not small men, moved by jealousy. They were puzzled men, who were groping their way through one of the most complex of all human relationships. Someone had failed to counsel them properly.

It is not easy for the average young captain, who by reason of his youth is usually somewhat lacking in self-assurance and in the confidence that he can command respect, to understand that as a commander he can grow in strength in the measure that he succeeds in developing the latent strength of his subordinates, or having accepted this idea as a principle, to apply it as his rule of action.

But for that matter, the senior commander who cannot learn to function through a staff, and get the most out of himself and his men by so doing, is by no means a rarity.

6

FIRE AS THE CURE

"Finally, it is the volume of fire that counts. You win if you can kill more of the enemy than he can kill of you. If you cannot, you are defeated."

—Secretary of War Robert P. Patterson.

SINCE it is axiomatic that the enemy is defeated finally by the fire which beats him back and the movement which makes his displacement permanent, it is elementary that in all operations those soldiers which advance but do not willingly employ their weapons still make a direct physical contribution to the success of offensive action.

The effect of their presence in the zone of fire is stimulating to their comrades and even may be depressing to the morale of the enemy if it becomes revealed by his reconnaissance.

Soldiers along a fire line are singularly unaware of whether the number of their comrades who are actually firing is large or small. Under the best of circumstances they cannot "feel" what is going on very far toward either flank. Too, they are usually preoccupied with their personal tactical situation to a degree which shuts out any organized impression of what others are doing with their weapons. In dealing with several thousand squad and platoon leaders after battle, I have yet to meet the man who could say what any one of his group had done with his weapon for so long as an hour.

This should be accepted as a fact and as a basic principle: The moral feeling of physical support in battle derives from the presence of another soldier rather than from the knowledge that he is taking appropriate action.

The inaction of the passive individuals does not have a demoralizing effect on those who are making tactical use of their fire power. To the contrary, the presence of the former enable the latter to keep going. Every potential effective along the line who is within sight of any other soldier adds moral strength to the line. It is only when men begin to give ground that courage wavers all along the line. And while it is clear beyond challenge that the true defensive strength of the position is in those men who use their weapons, there is no proof that the soldier who will not take the initiative in firing against the enemy will quit the ground any sooner, under pressure, than his more aggressive comrade. We are ever prone to lump the military virtues together and to take it for granted that voluntary action is the outward sign of the courage to face death bravely and that the absence of it is a sign of unheroic qualities in the individual. But these are ill-reasoned conclusions.

Simply to release man from the fear of death does not insure that he will act as if he were immortal.

In the Pacific campaigns our forces were impressed time after time by the phenomenon of enemy troops who had quit their arms and who appeared incapable of any offensive or self-protecting gesture. Yet these troops stood their ground like plants rooted in the earth and insisted on being killed to the last man. Their living bodies were the defensive base around which the action of their more willing comrades proceeded.

We should take it that the initiative to fire is only one positive quality in the good soldier. Notwithstanding that it is the mainspring of successful minor tactics and hence of final victory in war, those who are incapable of developing it are not to be too greatly discounted for this one fault.

In the equating of tactics these passive soldiers may be considered as contributing their weight to the mass of the attack but little or nothing to its velocity. For it is clear to be seen that if none were to fire, movement alone could not overcome the fire of the enemy, either by inflicting casualities upon him or by frightening him and forcing him to yield ground. Fundamentally, fire must always be beaten by fire. Fundamentally, movement is the means of increasing the efficiency of one's own fire until at last the strength of the enemy's fire is reduced to the vanishing point.

The rarest thing in all battle is fire in good volume, accurately delivered and steadily maintained. And yet, until the day of final surrender, the far purpose of all movement in war is the upbuilding of one's own fire power and position at the expense of the enemy's power and position. Any movement which is not concentric with this purpose is contrary to the principle of conservation of force and is therefore wrong. This applies equally to a retreat or to a pursuit. "Let us not think of mobility in an army," General Charles P. Summerall once said, "unless we think of mobility accompanied by violence." General George S. Patton frequently expressed this same idea. But what is violence, as the word is used by these two commanders? It is simply fire—fire in all of its forms. Nothing need be added to that definition and nothing should be taken away.

So it is a curious thing that even in professional circles there is a constant obscuring of the main idea that fundamentally fire wins wars and that every other aspect of operation is important only in the measure that it contributes to this grand object.

The familiar words of Napoleon: "It is on supply that war is made," simply underscores the fact that primarily war is made *with* fire, and that logistics have a decisive effect upon the arena only when they enable military forces to bring a superior fire to bear. Likewise with the equation, said by Napoleon and repeated by Foch: "In war the moral is to the material as three to one." This is a truth only as it is related to the state and possibilities of fire. Among fighting men morale endures only so long as the chance remains that ultimately their weapons will deal greater death or fear of death to the enemy. When that chance dies, morale dies and defeat occurs.

But armies from well-civilized states are so strongly influenced by civilian thinking that in their desire to refrain from circulating any ideas which may be shocking to civilian sensibilities they sometimes slight their own first principles. That is one reason why the subject of fire is not given its just due. We are reluctant to admit that essentially war is the business of killing, though that is the simplest truth in the book. Indeed, it is so simple that many of the thinkers on war have passed it up in favor of half-truths which contain a greater element of mystery.

As an example, we can take that oft-quoted and misquoted statement of Nathan Bedford Forrest about "getting there first with the most" which is supposed to illuminate the vital tactical principle. This apotheosizing of mass and mobility is so false as to constitute a dangerous illusion. It is wholly at odds with the real secret of mobility as it was understood by the Great Captains or even as it was practised by Forrest. They moved faster because they could place their trust in the superior hitting power of relatively small forces. It is not coincidence solely that the commanders who are renowned for their speed of movement were also the masters of the application or fire. The essence of success in tactics comes of what you do with fire after you get there. In his realistic restatement of Forrest's principle, Major General Charles W. O'Daniel put it this way: "In battle, it is a matter of getting there first, regardless, and then having the ability to stay put."

For the infantry soldier the great lesson of minor tactics in our time, which is at the same time the outstanding moral to be drawn from study of the "small picture" in this last war, is the overpowering effect of relatively small amounts of fire when delivered from the right ground at the right hour. The mass was there, somewhere in support, and mobility was needed to put the vital element in the right place. But the salient characteristic of most of our great victories (and a few of our defeats) was that they pivoted on the fire action of a few men.

In the whole of the initial assault landings on the Omaha Beachhead, there were only about five infantry companies which were tactically effective during the greater part of June 6, 1944. In these particular companies an average of about one fifth of the men fired their weapons during the day-long advance from the water's edge to the first tier of villages inland—a total of perhaps not more than 450 men firing consistently with infantry weapons in the decisive companies. These facts were determined by a systematic check of the survivors. It was not a story of great volume, even for the men who fired. Only one company was able to unite a base of fire for any period. The company which made the deepest penetration, losing a high percentage of its men in so doing, saw only six "live Germans" during its advance, and these turned out to be Russians. The day was conspicuous for its lack of live targets.

Yet had not this relatively small amount of fire been delivered by these men, the decisive companies would have made no advance in their separate sectors, the beachhead would not have begun to take form, and in all probability Normandy would have been lost. At their backs was the power of the mightiest sea and air forces ever to support an invading army in the history of the world. But in the hour of crisis for these infantry companies, the metal, guns and bombs of these distant supporters were not worth three squads from that small band of men which had gone to work with their grenades and rifles.

These riflemen did not win the victory at Omaha Beach. To say that they did would be giving them too much credit. But without them, there would have been no beachhead and no victory.

Again, could there be a better example of the miraculous possibilities of a very small volume of fire than the incident at the Bourcy road block to the north of Bastogne on the morning of December 19, 1944? Twelve American armored infantrymen—twelve very nervous infantrymen—fired erratically in the darkness at a group of approaching enemy soldiers. They fired and fell back. They were looking for better ground. They thought they had probably turned back a German reconnaissance element and that their fire may have hit four or five men.

But the German group was the point of an infantry regiment which was leading the column of the 2nd Panzer Division. It had recoiled on meeting the surprise fire. The commander reported, quite incorrectly, that he was being opposed by superior forces. The word was passed through two higher headquarters and Corps ordered 2nd Panzer Division to change its route of march and swing northward, thereby wasting precious time and traversing unnecessary space. Had the enemy made one good lunge against the Bourcy road block, he could have turned southward and entered Bastogne before the American forces had assembled.

The whole body of evidence from our own and enemy sources supports the conclusion that had this happened, the Ardennes campaign would have run a far different course and the enemy would not have been checked short of the line of the Meuse.

One other example: Because a few paratroopers were dropped far off course, and though unorganized, had the courage to address the enemy with fire around Le Ham and Montebourg in the early morning hours of June 6, 1944, the German High Command concluded that this marked the nothern boundary of the planned American effort. As a result, troops were held to the north of Montebourg through the greater part of that day, which troops might have broken the 82nd Division's tenuous hold on Ste. Mère Eglise, had they attacked southward at once.

If I am correct in the proposition that the decisive importance of fire is beyond exaggeration, it must follow that at all levels of command the intensifying and directing of fire becomes the foremost problem. As far rearward as the bureaus of the War Department the aim should be to get the utmost impact into hitting in the areas where the enemy is hittable. It is the true aim of training and of indoctrination; lines of communication exist but to serve it; all other goals of the national defense are subordinate to it.

Yet one curious contrast is to be noted. In the rear areas and as far forward as the no-man's land which is dominated by the bullet, the attack upon the basic problem is in the main logistical. Along the line where the company commander takes over, however, the increasing of fire volume must be considered primarily as a psychological matter. Only so can the commander get the highest level of combat efficiency out of his company.

Since the average man likes to fire a weapon and takes unreluctantly to instruction on the range, it cannot be doubted that a majority would participate freely with their weapons under conditions approximating a field exercise.

But combat cannot ever approximate the conditions of field maneuvers. Fears of varying sort afflict the soldier in battle. The unit commander soon comes to realize that one of his difficulties is to get men to leave cover because of enemy bullets and the fear they instill. In training, there being no real bullet danger even on the courses which employ live ammunition, every advance under a supposed enemy fire is unrealistic. Too, in training, the soldier does not have a man as his target. He is not shooting with the idea of killing. There is a third vast difference in the two conditions: The rifleman in training is usually under close observation and the chief pressure upon him is to give satisfaction to his superior, whereas the rifleman engaging the enemy is of necessity pretty much on his own, and the chief pressure on him is to remain alive, if possible.

When the infantryman's mind is gripped by fear, his body is captured by inertia, which is fear's Siamese twin. "In an attack half of the men on a firing line are in terror and the other half are unnerved." So wrote Major General J. F. C. Fuller when a young captain. The failure of the average soldier to fire is not in the main due to conscious recognition of the fact that the act of firing may entail increased exposure. It is a result of a paralysis which comes of varying fears. The man afraid wants to do nothing; indeed, he does not care even to think of taking action.

Getting him on his way to the doing of one positive act—the digging of a foxhole or the administering of first-aid to a comrade—persuading him to make any constructive use of his muscle power, and especially putting him at a job which he can share with other men, may become the first step toward getting him to make appropriate use of his weapons under combat conditions. Action is the great steadying force.

It helps clear the brain. The man who finds that he can still control his muscles will shortly begin to use them. But if he is to make a rapid and complete recovery, he requires help from others.

In the attack along the Carentan Causeway during the night of June 10, 1944, one battalion of the 502nd Parachute Infantry was strung out along a narrow defile which was totally devoid of cover and where throughout the night the men were fully exposed to enemy bullet-fire from positions along a low ridge directly in front of them. The ridge was wholly within their view and running off at a slight angle from the line of advance of the column, so that the Americans were strung out anywhere from 300 to 700 yards from the enemy fire positions.

In this situation the commander, Lieutenant Colonel Robert G. Cole (later killed in action in Holland) was able to keep moving up and down along the column despite a harassing fire, and observe the attitude of all riflemen and weapons men. This was his testimony, given in the presence of the assembled battalion: "I found no way to make them continue fire. Not one man in twenty-five voluntarily used his weapon. There was no cover; they could not dig in. Therefore their only protection was to continue a fire which would make the enemy keep his head down. They had been taught this principle in training. They all knew it very well. But they could not force themselves to act upon it. When I ordered the men who were right around me to fire, they did so. But the moment I passed on, they quit. I walked up and down the line yelling, 'God damn it! Start shooting!' But it did little good. They fired only while I watched them or while some other officer stood over them."

In the early light of the following morning the battalion closed with the enemy and drove him back from his fire positions along the low ridge. There ensued a day-long battle, marked by the closest kind of fighting, as the enemy came on five times in counterattack along the hedgerows, trying to regain the initial position.

In the crises of these actions the two forces were about 40 feet apart. The fight had opened with a bayonet charge and about six of the enemy had been killed with that weapon. In local skirmishes the machine gunners were finding targets at less than 20-yard range. The day ended in complete victory for the American side at a cost of about 40 per cent loss in strength. Throughout the day the tactical situation had been as deadly as it was fluid and there was always plenty of blank space along the hedgerows for any man who was willing to engage. The ammunition supply kept coming.

Such was the persistence of the enemy in his effort to advance along the ditches and hedgerows that the day produced an exceptional number of live targets. In fact, we found by careful check of the battalion that exactly one fifth of the battalion had seen one or more live enemy soldiers at some time during the battle. But we found further that, even so, only one man in four among the Americans had employed a weapon of any kind. Also, the circumstances of the tactical action established that the majority of casualties on the American side had come from mortar and artillery fire while at varying distances to rear of the front line of resistance, which would appear to warrant the conclusion that the ratio of active firers would not have been increased had it been possible to poll the dead and the critically wounded.

I followed this same battalion through the airborne invasion of Holland in September, 1944, and through the winter fighting in the Ardennes, and I doubt that there has ever been a finer fighting unit in the Army of the United States. It never tasted defeat nor was it ever given an easy assignment. At least three of its engagements are historically noteworthy examples of heroically successful achievement against great odds. It was tested over marshland and through hedgerow country. In Holland, west of Zon and near the Wilhelmina Canal, its hardest engagement was fought through a checkered pine forest on flat ground; the enemy had enfiladed every forest trail with machine guns and from the other flank and from the front his artillery kept the woods under a point-blank fire. Perhaps the battalion's finest hour was had on the rolling hills northwest of Bastogne during the early stage of the defense of that town in December, 1944. But while the battalion matured constantly from battle experience, there was no marked percentage increase in the number of men who used their weapons to fire at the enemy. The figure was as I have already quoted it—between 25 and 30 per cent. Even so, it had as high an efficiency of fire as any unit which I have ever known.

Prince Hohenlohe was profoundly right when he said: "It is proof of a superior military instruction if in battle the men only bring their rifles up to their shoulders to fire."

While it is a natural and pardonable fault in the average commander who has had the good fortune to lead troops in battle that he is ever after disposed to make his men appear more resolute than they are by nature, it is a tendency which thwarts realism in the training of men for war. Would it not be better for the unit and for the army if the commander knew from the beginning that under the usual conditions of engaging the enemy with small arms fire, only a minority of his riflemen are likely to fire freely and voluntarily? Once this is established as a factor in his mind, and he has been given the offsetting assurance that the fire line opposing him will probably operate under the same handicap, he will be more prone to consider the ratio of fire within his command as a special subject for his battlefield observation and correction.

The diagnosis of the disease must precede the remedy; the

object of search must be known before there can be intelligent seeking.

Once oriented toward the problem, the commander would work more carefully with his junior leaders in training. They would be told that in combat it would become one of their tasks to mark well the men who take the initiative with the rifle or other hand weapons. When these men are identified, it then becomes incumbent on the junior leader to devote more of his effort to personal work among the nonstarters, encouraging them to work up to favorable fire positions and giving them direct orders to begin fire with the weapon.

This will not produce a cure-all but it is at least a start. The survey of the company and an organized knowledge of how its individuals respond under actual fire conditions should precede all else; correction of this study should continue thereafter.

When the tendency of all members of the unit has been thus appraised, the commander may find that for maximum combat effectiveness it is necessary to make a number of personnel adjustments. Men previously overlooked will be marked for promotion. The man who is always a self-starter under fire is not *ipso facto* qualified for junior leadership; there must be other substantial elements in his character. But on the other hand, the NCO who cannot exercise fire initiative will lose the respect of his men as quickly as his weakness is observed by them in battle. Even the soldier who cannot overcome a similar weakness in himself will look with contempt on a superior who appears to shirk his duty because of danger.

The men who show no disposition to use the small weapons, even when properly urged and directed, can be switched to the gun crews. There, the group will keep them going. Men working in groups or in teams do not have the same tendency to default of fire as do single riflemen. This is such a well-fixed principle in human nature that one very rarely sees a gun go out of action simply because the opposing fire is too close.

As another experiment, unwilling riflemen may be switched to heavier and more decisive one-man weapons. This sounds like a paradox—to expect greater response to come from increased responsibility. But it works. I have seen many cases where men who had funked it badly with a rifle responded heroically when given a flame-thrower or BAR. Self-pride and the ego are the touchstone of most of these remarkable conversions. A man may fail with the rifle because he feels anonymous and believes that nothing important is being asked of him. (Though that is a false feeling and the rifle must remain the prime weapon of the infantryman, I remind the reader that we are considering human nature in its relation to weapon efficiency.) The switch to a heavier weapon is a challenging form of recognition. It is a chance for the man to show others that he has been held in too lowly esteem.

Whether there is any sure cure with the rifle itself I am not at all certain—whether it would be possible by special techniques to break down a rifleman's resistance against employing his weapon upon human targets. To think that the job can be done simply by giving the man confidence in his weapon or working him up to the point where he enjoys firing it is a gross miscalculation. These things are a valuable part of the conditoning process but they will not remove the final mental block.

To my knowledge, no sustained experiments have ever been made during combat to see whether and how a group of non-firers can be converted to willing use of the rifle. Moreover, I doubt that a test under non-combat conditions would have value. The only thing that counts is how the man responds when he is given opportunity to fire at an object for the direct purpose of taking another man's life. Let me cite an example:

> In the 184th Infantry Regiment's sector during the Kwajalein battle, we saw two objects floating by, 200 yards out in the lagoon. They looked like the heads of swimming men. From forward of us, there was a spattering of fire which kicked up the water around the objects. The riflemen close around me—there were about ten of them—held their fire. I then turned my field glasses over to them, saying: "Take a look and you will see that those men are wasting their ammunition on blocks of wood." They did so, and within a few seconds they were all firing like mad at the objects. They had found a release in the very information which I had supposed would cause them to hold their fire.

There were numerous incidents in this battle and in others wherein enemy soldiers walked deliberately into the open in full daylight, exposing themselves long enough for a score or more men to get a sustained view of the target, without one shot being fired.

As to the main problem, I suspect, and my psychologist friends assure me that it is so, that if any treatment is likely to work it would be to dispose the rifleman where he has a clear sight of an enemy target, then handle him as one would a recruit on the range, making certain that he continues the fire for an extended period. By that, I do not mean necessarily a live target. They are not that convenient. The instructor who has his pupil hold fire until he sees a man to fire on will usually have a very long wait indeed. But it is necessary that the fire be aimed against a position where the enemy is presumed to be located.

It seems reasonable to believe that there is a definite advantage to getting the soldier into the habit of free firing in combat while the situation is still such that his target is a position rather than a man moving clear. It becomes easier for the supervisor to regulate his own work and it is the logical and most promising approach to the problem, if only for the reason that the average firer will have less resistance to firing on a house or a tree than upon a human being. To clear up this point, it is necessary to take a somewhat closer look at the average, normal man who is fitted into the uniform of an American ground soldier.

He is what his home, his religion, his schooling, and the moral code and ideals of his society have made him. The Army cannot unmake him. It must reckon with the fact that he comes from a civilization in which aggression, connected with the taking of life, is prohibited and unacceptable. The teaching and the ideals of that civilization are against killing, against taking advantage. The fear of aggression has been expressed to him so strongly and absorbed by him so deeply and pervadingly—practically with his mother's milk—that it is part of the normal man's emotional make-up. This is his great handicap when he enters combat. It stays his trigger finger even though he is hardly conscious that it is a restraint upon him. Because it is an emotional and not an intellectual handicap, it is not removable by intellectual reasoning, such as: "Kill or be killed."

Line commanders pay little attention to the true nature of this mental block. They take it more or less for granted that if the man is put on such easy terms with his weapon in training that he "loves to fire," this is the main step toward surmounting the general difficulty. But it isn't as easy as that. A revealing light is thrown on this subject through the studies by Medical Corps psychiatrists of the combat fatigue cases in the European Theater. They found that fear of killing, rather than fear of being killed, was the most common cause of battle failure in the individual, and that fear of failure ran a strong second.

It is therefore reasonable to believe that the average and normally healthy individual—the man who can endure the mental and physical stresses of combat—still has such an inner and usually unrealized resistance toward killing a fellow man that he will not of his own volition take life if it is possible to turn away from that responsibility. Though it is improbable that he may ever analyze his own feelings so searchingly as to know what is stopping his own hand, his hand is nonetheless stopped. At the vital point, he becomes a conscientious objector, unknowing. That is something to the American credit. But it is likewise something which needs to be analyzed and understood if we are to prevail against it in the interests of battle efficiency. I well recall that in World War I the great sense of relief that came to troops when they were passed to a quiet sector such as the old Toul front was due not so much to the realization that things were safer there as to the blessed knowledge that for a time they were not under the compulsion to take life. "Let 'em go; we'll get 'em some other time," was the remark frequently made when the enemy grew careless and offered himself as a target.

To get back to my main point, however, it would likewise seem reasonable to believe that if resistance to the idea of firing can be overcome for a period, it can be defeated permanently. Once the plunge is made, the water seems less forbidding. As with every other duty in life, it is made easier by virtue of the fact that a man may say to himself: "I have done it once. I can do it again."

As for those other men, the self-starting men who somehow have managed to overcome their inhibitions and have proved their initiative under fire beyond all doubt, the good commander will cherish and protect them as if his own life depends on it, for surely his professional reputation does. However, it is self-evident that for the good of the company such men should not be wasted on rear area or communications duty unless the signs of cracking from battle strain become evident. When that happens, arranging some special assignment which will afford the good soldier temporary relief is the commander's obligation. On patrol, outpost, or other hazardous duty, the commander would be ill-advised to concentrate these moral leaders of the company, though they should be present in the minimum numbers which will provide a safe binder for the other files so assigned. Too, he will take care that they are never driven by his other subordinates; they do not require driving and there can be no surer way to destroy their mettle.

Finally, when the rewards of battle are handed out, he will make certain that these are the men who are honored first. In my judgment the soldier who consistently addresses the enemy with fire is full worthy of decoration. But what honors are commonly given such men under our present awards system? None whatever! It is the almost universal practice of boards sitting in judgment somewhere safely in the rear areas (I do not speak from inexperience, having sat on four such boards) to dismiss all such cases with the comment: "He was only doing his duty." Until it is formally recognized that there could be no higher tribute than this to the combat soldier, our system of distributing awards will tend toward the discouragement of the fighting line rather than otherwise.

The figures quoted through this chapter may be startling. But I see no reason to believe that they will alarm those who have studied the reality of combat rather than the romance which has been written about it. Though great strength may be present and its preponderance may decide the fortunes of the day, the decisive local and tactical issues on the field of fire are invariably decided by very minor forces.

A regiment may be committed. Of that regiment, one

battalion advances to small arms fire contact with the enemy. Finally, from that battalion, two depleted platoons get to the outwork and three squads make the last run which clears the enemy from the final ditch. Thus the course of an average action.

Far from discouraging the young officer, some reflection and emphasis on these aspects of the infantry fire problem will give him fresh inspiration for study of his men and for doing patient personal work among them. It is a challenging fact that the persuading of only a few more rifles beforehand may give him a favorable tactical balance.

The broad inference to training is that it is unprofitable in general work with the rifle to put the accent on live targets or even on clearly defined targets such as those used for record. The moral effect on the rifleman is almost paralyzing when he moves from these stereotypes to a battlefield where he is told to open fire on some apparently innocent feature of the landscape. Indeed, so much was said in training for the past war about harboring ammunition and making certain of the target that it became a brake upon field operations. The ranks frequently objected that their officers were overriding their own principles when the time came in battle when they insisted on heavy fire with no targets to be seen.

Undue emphasis on conservation is as great a danger to fire power as is an excess expenditure of ammunition. Bullets kept in the magazine when they should be fired are certainly bullets thrown away.

The prime object is to insure by training that men will fire when ordered. To get that result without lessening the accent on marksmanship, we should put a great deal more on musketry whereby the formations will acquire the habit of massing fire whenever ordered and against whatever target may be designated—the embankment of a river, the bases of the forward trees in a line of woods, or the crest of a hill.

We need to free the rifleman's mind with respect to the nature of targets. This sounds simple. But the requisite flexibility cannot be obtained by such fire exercises as sending men against a "Little Tokyo" on the edge of a reservation or through what they learn on maneuvers, where there are usually plenty of live targets and there is a tendency to restrict fire until the live targets are observed. The proper educating of group fire requires constant insistence on the principle of spontaneous action developing out of a fresh and unexpected situation.

Once the system of free selection of targets is installed, the furthering of a high state of fire morale will proceed according to the thoroughness of control by the junior leaders. What that entails was once well stated by Colonel de Grandmaison:

> As regards fire leaders, teach them first how to regulate their groups; secondly, how to guide their fire. Explain to them that they will control the fire of their men in proportion as their orders are simple and necessary. Teach them not to regulate the fire but to regulate the firers. Teach them that it is more important to place their men in good fire positions than to issue correct orders.

I would add this: "Teach them that in battle many of their men will not fire unless given a specific order and that their first responsibility is to mark these men and devote personal attention to them."

The doctrine of fire discipline has accented for so long the need of controlled fire that it has almost obscured the fact that the fundamental problem is how to build up fire volume and develop more willing firers. One cannot deny that looseness of fire at times creates a certain hazard for troops. But this problem must be viewed in proportion; we cannot afford to miss the forest for the sake of a few trees. Though it is hard on the nerves at the time, so far as the end result is concerned, it is better by far to have a company of green, trigger-happy soldiers than a company which lacks the will to use its weapons. The former will make a recovery from nervousness as they become more accustomed to the sights and sounds of combat and the tense silence of the lulls in between fighting; the latter will never be given the chance.

And last, if we are to strengthen sound training principles and establish mental attitudes which are essential to the understanding of the decisive importance of fire in tactics, we will be well advised to cease talking about "fire and movement" as if the latter were separate and apart from the former in tactical fact, and there did not exist an automatic and unbreakable connection between them.

What then should replace the emphasis on fire and movement? Simply this—a basic understanding of the reality that fire superiority is the thing and that movement is its physical and psychological derivative, along with all other acts of the initiative.

I believe this to be one of the simplest truths of offensive power and the basis of sound minor tactics. But it is so simple that it has been largely overlooked while we have sought for some more complex answer on how to get velocity in the attack. *The soldier who learns and applies correct principles of fire will always move.* He has solved the real problem, the problem which every unit commander of combat forces in World War II will say was his main problem. By so doing, he comes into adjustment with his tactical situation. It is the other fellow—the man who will not apply fire—who provides the brake on movement. The man who has the fire habit is looking always for forward ground from which to give his fire increased effectiveness. This is as inevitable as the working of the law of gravity.

As with the man, so also with the unit. The tactical body which attains to fire superiority by its own sheer merit will never fail for lack of mobility unless it is wholly let down by the forces in the rear area. Fire is the key to mobility. To fire is to move. Weapons when correctly used will invariably bring decision.

But without superior fire power, mass and velocity can never win a war.

7

THE MULTIPLES OF INFORMATION

"I consider to be of utmost importance the keeping informed of our men as to current situation so that each man may perform his duty with understanding of its importance."

—Sergeant David Thibault, in a letter written from the African Theater.

HAVING started to call this chapter "Communications," I decided against it for fear that use of that rather formidable word might interfere with my communicating to the reader what I regard as a vital but frequently overlooked principle of minor tactics.

If a word comes to mean too many things, it frequently misses fire at the critical point. That has been the fate of the word "communications." It has become another military catchall, and because it means so many things, it quite frequently means nothing.

To illustrate: A regimental commander asks a company commander: "How are your communications?" and the latter replies: "Excellent!" because his telephone line is working and his supply is coming forward. And at the same time he is at the point of despair because he hasn't had any worthwhile intelligence from his flanks for hours! An exaggeration? Not at all. I have seen it happen on more than one occasion.

As another example, I would like to point out that in practical fact the two words "communications" and "contact" are inseparably linked in their tactical application and that each flourishes according to the strength of the other—and here, of course, I speak of friendly contact. Yet they are not so considered by the majority of younger commanders during combat, and it is only after long experience in battle that they come to see that contact without communications is like fighting in the dark when it is possible to switch on the light. Therein lies a great weakness.

One would think that it would become almost second nature to the commander to reach eagerly toward supporting forces. One would imagine that whenever on the field of battle he made contact with a friendly element, that fact would flash a red light in his brain, causing him automatically to raise the questions: "Have I established full communications? Do I know the strength and intentions of the force now helping me? Do they know my strength and intentions?"

But indeed, such is not the case. Not one officer in two score is inclined to make this a rule of action; it is the rare man who applies it even on occasion.

Yet one who pursues the study of operations at the lower tactical levels must reach the ineluctable conclusion that failure to grasp this simple and basic idea of communications is the most common cause of breaking of contact on the field of battle and is the general reason why the union of tactical forces is so frequently deprived of any positive result.

The tendency of the minor commander is ever to do more worrying, but less acting, with respect to his flanks than with respect to his front and rear. Encompassed by problems, he is inclined to discount the importance of his personal action toward left and right, or to think of his flanks primarily as props to his own position rather than as co-equal forces in a mutual undertaking.

What he does not see is that the juncture of forces calls for a juncture of thinking and of information if success is to be had at the lowest cost. The total strength of a position does not reside in its numbers of men and weapons but in knowledge of the numbers and the mutual sharing thereof. All tactical support must be known and be felt to be of true moral help in a time of crisis. That part of it which lies beyond the knowledge of the ranks of a company—the supporting artillery fire which it cannot see or the strong point lying just around the bend in the river—may be greatly sustaining to the company's efforts in terms of protection to front and flanks or actual hurt to the body of the enemy, but so long as it remains unknown, it will not keep the company from breaking when the pressure appears to become uncontainable.

Let us take this situation. A company is defending a bridgehead. It is supported on both flanks by other com-

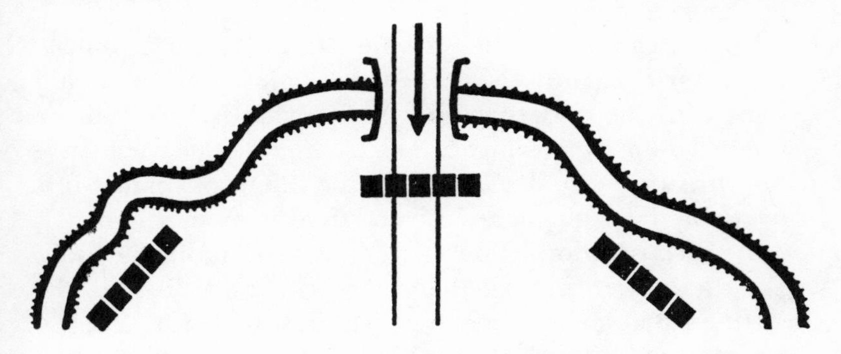

panies. But these companies are around the bends in the river. The members of the defending company do not know of them because they cannot see them and have not been given complete information. The enemy pressure from across the river intensifies until the defending company sees that its own flanks are in jeopardy. It then breaks back and by so doing compels the abandonment of the entire position. It need not have happened, since a common sharing of information about strength would have given the company assurance that the pressure was still containable by the battalion. Lest it be thought that this is pure hypothesis, I will add that I have seen ground lost in just this way on perhaps six occasions by the American Army.

So it should be a watchword of minor tactics that it is never enough to support to the limit; tactical security requires that full knowledge of all support be shared by all concerned. Better that a company know that it is being helped by a depleted platoon on its left, so long as that knowledge is mutual, than that it be supported on its left by a full company, with neither knowing. Nor can this kind of knowledge be imparted simply with a wave of the hand if it is to have true value. If men are to believe, they must have specific information—just where the others are and what they propose doing.

The search for this kind of information is a prime obligation of the company officer in combat. Moreover, he should begin with the understanding that very little of it will be supplied him by the higher headquarters. They will give him what they can, but under the conditions of combat it is invariably inadequate, and by the time it reaches him it is cold. Lack of information as to the situation on the flanks is of itself a warning signal and should become likewise the starting signal for someone to be sent in search of it. Yet this is a responsibility which one finds otherwise efficient company officers most reluctant to assume. When hearing nothing from left and right, our average combat officer is much more likely to spend some time roundly abusing the negligence of these other forces than to heed the red light which cautions him that the flanks must be equally in the dark about the situation in the center and that therefore someone must take the initiative or all will remain uninformed. However it is a not altogether unpardonable weakness as nothing on the field appears to entail more mortal danger than hunting around for a loose flank, though I rather believe that the greater cause is that during training an officer becomes too closely wedded to his rearward communications.

Rare, indeed, are the circumstances in modern war when a man can see, hear, or feel the strength of supporting elements for any great length along a line of battle. The nature of the terrain over which maneuver forces proceed toward engagement, the nature of protection, and the physical reaction to hostile fire all determine that forces which are endeavoring to remain invisible to the enemy must remain largely invisible to their own components.

The common scene in open warfare is a landscape; the total absence of moving things is the surest sign that one has reached the line of danger. But many men have failed to recognize or heed that warning and have simply blundered into death. Even when movement ceases and the opposing fire lines again become static, a reconnaissance along the friendly line is a point-to-point search for the hideouts of men, which is largely fruitless unless it is done by map. One not knowing where to search might move for miles along a main resistance line and see hardly a sign of war. On the battlefield it is only when the fire impasse has been at last broken that military forces have that appearance of continuity of strength in line which the uninitiated mistake for the reality of a victorious tactical formation when they see it on the motion-picture screen.

In combat almost nothing has the appearance of juncture and of hanging together. Viewed from above, an attack would appear not unlike the disparate movements of a colony of water bugs. The first effect of fire is to dissolve all appearance of order. This is the most shocking surprise to troops who are experiencing combat for the first time. They cannot anticipate the speed with which their own forces become fractionalized or the extent to which the fractions will become physically divorced from each other as the movement is extended and enemy resistance stiffens.

During the Normandy fighting there was much emphasis on the ill effect of the bocage country in compelling a rapid breakdown of the smaller tactical units, thus compounding the difficulties of control. But this was no new problem in tactics. The main difference was that the hedgerows and their effect on a formation were fully visible to the naked eye. It was easy to see what was happening and why.

But a comparable effect is produced by almost any terrain which can serve infantry forces, including most desert country. It is not the accident of ground which produces the effect but the simple fact that men must take advantage of the accident in order to survive. House-to-house fighting in a town or city (and regardless of what the book says, this is always a catch-as-catch-can business) will split a company apart more quickly than any other kind of action. The hedgerows notwithstanding, in Normandy it was relatively easier for forces to maintain contact among their own elements than in the campaigns occurring at the same time in the Central Pacific where the troops were advancing across flat, palm-covered islands.

The chief features of ground in the atoll fighting, as revealed by air reconnaissance prior to invasion, were the shallow mangrove swamps and bobai pits, a relatively open vegetation, the native huts, and the enemy's defensive works. But in the first stage of invasion the shore bombardment and air bombings worked these places over until most of their features were battered beyond recognition; the preliminary fires of the shore-based artillery completed this ruin. Trench systems and roads were erased. Surface structures looked as if they had been up-ended by a typhoon. Consequently, maps meant little and all movement had to proceed by eye contact if juncture was to be kept. Yet because it was physically impossible for troops to advance very far without some parts of the line having to break and detour around some impassable object, the effort to keep contact by sight was thwarted many times. In the island fighting there was simply no system of liaison known to man which could have beaten the natural obstacles. That was not quite true of Normandy; the geometrical design of the countryside did not make contact impossible though it did multiply the difficulties.

Even in extended linear works such as the Siegfried Line or the great trench systems of World War I there is no magic link uniting the moral strength of troops. They cannot see what goes on very far beyond the first traverse in the trench or the first hump of ground intervening between two emplacements. Consequently, such support as they derive from their flanks, as a moral value, must be dependent upon the flow of information.

I think the direct connection between informational strength and weapon strength as the complementary halves of moral strength has seldom been more graphically illustrated than by the story of the personal experience of Lieutenant General Herman Fritz Bayerlein, commander of Panzer Lehr Division (*Bastogne: The First Eight Days*). He is already in a state of alarm, having misinterpreted the events of the morning and misestimated a situation which was greatly in his favor had he but known it. He hears heavy fire far over on his left, near the village of Wardin. It is the victorious fire of his own Reconnaissance Battalion which had half-destroyed and scattered the American company which had held the village briefly. But Bayerlein does not know that his own forces had attained to this forward ground in any strength. So he concludes that the heavy fire comes from American forces entering the village. Certain that he has been outflanked, he prepares to withdraw his center and recommends to the Corps Commander that the attack on Bastogne be suspended until he can extricate his Division from an impossible position.

Information is the soul of morale in combat and the balancing force in successful tactics. Yet in an era of warfare which is on the whole extremely enlightened, when we are so concerned for the welfare of troops that we strain our supply lines so that fresh eggs and oranges may be served in the front line during the course of the most rapid advance by field armies in history (Germany, April–May, 1945), we have not found the means to assure an abundant flow of that most vital of all combat commodities—information.

It is true that we have worked marvels in furthering the rearward flow of information to higher headquarters. When a small and highly mobile force of men seized the bridge at Remagen the fact was known to the Supreme Commander, then at Rheims, within the hour. What happens in a line company is quite likely to become known to the staff at a corps headquarters 20 miles to the rear within the space of a few minutes. The rub comes of this—that in all probability it will not become known to other companies within that same battalion in the course of the same day, if at all. Yet these are the people to whom the information would be most useful.

We can look briefly at a few of the reasons for this pervading contradiction:

(1) There is lacking a general recognition of the supreme importance of the lateral flow of information.
(2) Command at the lower levels is too often neglectful of

the principle that it is not a channel of information only but a distribution point.

(3) Commanders at the lower levels tend to be the arbitrary judges of what information deriving from a source lower down would be highly useful to the other elements lower down instead of abiding by the rule: when in doubt, pass it along.

(4) Inertia.

But these are by no means the full explanation. We see in battle the recurrent phenomenon of command at its various levels hard-riding the lower quarters for information until each echelon becomes so bedeviled by this nagging and incessant pressure that the mere easing of it, by whatever device or strategem, is likely to become the primary concern of each commander in his turn. He has time to think of little else. The satisfying of the higher level becomes the main object of operations.

It is the worst vice in operations and it is no respecter of persons; the victim is as likely to be a division commander as the leader of a platoon. But when it strikes through to the lowest level its consequences become intolerable and one finds it difficult to excuse the offender with the observation that the fault is in human nature. The dignity of command should require the curbing of that particular fault.

In the Pacific fighting I found company commanders joining a platoon in line just to isolate themselves from their telephones. They were literally "tired to death" of having the battalion commander insist on having a fresh progress report every fifteen or twenty minutes. And the battalion commander—poor devil—was only passing on the pressure which he had in turn received from a regimental commander who was trying to placate division. Yet one would observe that unless a battalion commander is wholly lacking in judgment or in intestinal fortitude, he should be strong enough to take this on his aching back and not pass it down to the headquarters which is at grips with the enemy.

In the Burton Island fight during the invasion of the Marshalls, one of these prodding demands for more progress raced from division right through lower headquarters to a platoon which had been stopped cold by Jap fire coming from spider holes arranged in great depth along the beach. Lieutenant B—— got the message and crawling forward to his most advanced rifleman, told him to get up and go on. The boy screamed: "So the whole god-damned Army wants to kill me, does it? O.K., Lieutenant, here I go, but watch what happens!" He was shot dead almost before he had gotten out of his tracks. That incident seared deep into the brain of every man who witnessed it. It was a final judgment on the futility of that kind of leading.

Now I would point out that the long-range jockeying of the lower echelons is not a genuine quest for information in the majority of cases. It is simply a search for mental easement via the telephone. It contributes little or nothing to understanding of the actual situation and it rarely contributes anything to tactical progress. In operations the object of any valid quest for information from the top down is to see what may be done to help. The application of a senseless pressure—and by that I mean a pressure which is at variance with the odds of the contest along the fighting line—serves only to destroy the confidence and wear the nerves of subordinates.

The all-too-frequent consequences of such pressure are lying, exaggeration, and distortion of the situation at the lower levels, resulting in a false concept of the situation at the higher. The average company commander can stand only a limited amount of this heat and then he will knock over a couple of outhouses and report that he has captured a village, or give the location of three cut-off and hopelessly placed riflemen as the approximate position of his left flank, even though he knows that his next move will be to withdraw them if possible. The effect is to make a wishful thinker of the most objective soldier. He reasons to himself: "I'll have that position in another hour, so I'll tell them that I have it now and get them off my back." Not infrequently this pseudo-optimism defeats its own purpose, for it gives the higher command a false idea of progress and keeps the company commander from getting the help that otherwise he might have received.

Since it is self-evident that the lower level can rarely have complete understanding of the problem of higher headquarters, it should be not less clear that information from front to rear will always be incomplete and indefinite unless the higher commands search forward actively for the data which they require for correct estimate and decision. This is true at all levels of operation. The higher commander cannot expect that the significance of the whole sweep of the landscape can be gathered for him by eyes which can focus on only a small part of the panorama.

The axiom that there is no substitute for personal reconnaissance applies as fully as ever, even though these are days when the majority of commanders and staff rely ever more heavily on the radio and the telephone.

One illustration will help to illuminate this proposition. An infantry battalion engages the enemy. The company on left reports that it is held all along the line. The center and right report "stubborn resistance" but steady progress until at last they are checked by battalion to preserve the integrity of the line. Now the company on left is not in position to advise battalion that the right and center are able to move because they are hitting the weak spots while the left is checked because it has bumped the enemy's local center of gravity. These are things that the company commander cannot know; indeed, the chances are 10 to 1 that he has heard almost nothing of the experience of the other companies.

Since none of the company commanders is likely to have qualified himself with information as to what is occurring along the general front, none can give a relative description of resistance in the several sectors. On the basis of what he hears by telephone, the battalion commander is therefore likely to conclude only that two companies are succeeding and one is failing. He will blame that company commander for falling down on his job and will put the pressure on him, drawing invidious comparisons with what is happening elsewhere along the battalion front. Once this idea becomes fixed, even a dynamite charge won't dislodge it—unless the battalion commander goes forward to see for himself or sends an influential member of his staff to take a new reckoning on the ground. Only then will it be recognized that what was being calculated wrongly as a local failure was in fact due to an initial deployment of the battalion strength which was based on a false assumption and that the now uncovered situation calls for a shifting of the battalion's center of gravity.

The rule applies equally to the relations between all higher headquarters which are directing tactical formations larger than the company and the battalion. Unless there is a constant boring toward the battle line by representatives of the higher command, it is most unlikely that the relative tactical situation—the state of pressure against one portion of the front as compared with another—can be truly clarified by wire or radio communciation. Men, even when given the same schooling, do not talk alike and see alike. Given the same degree of intelligence, they will still react according to character, the state of their nerves, or the knowledge acquired through experience. Under battle pressure the average commander tends to use such terms as "strong resistance," "intense fire," and "heavy shelling" in reporting his situation.

But these words rarely mean the same thing in any two situations or to any two commanders in the same unit. Therein lies a chief difficulty of getting that relative view of the situation which is requisite in well-balanced action.

There is the further point that higher headquarters needs to keep constant check of the understanding of its main tactical purpose at the lower levels, and that person-to-person contact is nearly an absolute requirement therein. A man leading a company or a battalion cannot be expected to know what importance Corps or Division attaches to the possession of a certain village or ridge, or to judge when the hour has been reached when, instead of trying to blow a certain bridge, it has come time to seize and defend it. Regardless of how sedulously he may be briefed in advance on these subjects, his own horizons must remain limited. Since he cannot know the flux and balance in the general situation, even the most soaring imagination will not enable him to see his small fight through the eyes of a general. It rarely happens in battle that the man who is actually winning or holding the decisive ground is privileged to see the significance of any part of it from the perspective of his Army commander. From this circumstance comes much missed opportunity. Consider the words of Major William R. Desobry, who commanded in the critical action at the village of Noville which meant the saving of Bastogne: "It was just another local affair. Not a man present had any idea of the importance of the engagement."

In Normandy, on D Day, one of the decisive bridgeheads across the Merderet River, by which the American forces would move westward to cut off the Cotentin Peninsula and seal the fate of Cherbourg, was held on four separate occasions by small American forces which then let it slip from their fingers. In two instances these forces marched away from the bridgehead to seek some lesser prize, leaving the bridgehead uncovered. In consequence, the progress of Corps was stalled for four days at the river crossing and victory was finally made certain only after bitter struggle and heavy loss.

These errors were not due to an absolute oversight on the part of any one person. As is usually the case in battle, things went wrong because of little slips all along the line. The small tactical forces which walked into the bridgehead and then walked out again had been several times briefed on the importance of the objective. But by their own testimony, the briefing had not given them a conviction that, above all else, they should make certain of the continued defense of the river passage.

In the event, they were thrown because they were too eager. It was Corps' misfortune that the attacking elements initially came into too easy possession of the prize, and because it was easy, they forgot that it was a prize and could not act according to what they had been told about its value. They were looking for a fight. When they walked onto the decisive ground they were not brought under fire. So they walked away, still looking for a fight. In the decisive moment, when thought was more important than action, they saw their tactical foreground through the eyes of small unit commanders rather than of a corps commander.

The question bears asking whether more than that might rightly have been expected of them. I doubt it. Such an expectation runs counter to human nature and to much of what we know about the flow of operations. It is one of the commonplaces of war that we see good troops fight bitterly for worthless ground which the enemy is strongly contesting and in the next round treat carelessly the really worthwhile object simply because the enemy momentarily does not seem to regard it as worth a contest. Since higher commanders are not above making that particular error, it is not remarkable that juniors sometimes fall victim of it.

That the Merderet incident involved airborne troops has an especial significance. For I believe it is incontestable that they had greater keenness than line infantry regiments, but that even so, all infantry troops of the first category are heading toward a form of war in which the greater number of engagements will be fully as complex as the situation which baffled so many men along the Merderet. The attack in future war will focus on the necessity for quick exploitation by tactical forces of a heavy initial shock and of the chaos which it creates. That means an ever-increasing use of airborne forces, whether paratroopers, glidermen, or regular infantry, moving by air transport.

The disarrangements which occur because of the lower levels not being able to see with the same eyes and think with the same brain as the higher command are a new and special danger which comes of widened deployments and the shock use of troops dropped suddenly onto decisive targets. It derives from a form of war in which initially there is likely to be little or no contact between friendly elements and control has to be built up from zero at the same time that the decisive engagements are getting under way. Though such situations were relatively rare even in World War II, they will be common enough in the next war.

For these reasons, the "principle of the object" will come to have many times its former importance in instruction to all ranks. The need for a clearer concept of it, however, is not greater than the need for junior commanders who will take a keen interest in the larger affairs of war and for higher commanders who make it a practice to get down to their troops. More appropriate to what we will know in the future than to what we have experienced in the past is that old truth: It is not always possible to lead from behind.

8

THE RIDDLE OF COMMAND

"Inquiries are now no longer made about customs that have been so long neglected, because in the midst of peace, war is looked upon as an object too distant to merit consideration. But former instances will convince us that the re-establishment of ancient discipline is by no means impossible, although now so totally lost."

—From VEGETIUS' writings on the Roman Legion.

THE flow of men and matériel during battle is ever toward the front. The flow of orders and instructions is toward the front.

But the prevailing flow of information, on which the employment of men and matériel in combat and the writing of orders and instructions for combat are based, is ever toward the rear, and the volume of it seems to increase according to the square of the distance from the fighting line.

The adequacy of information at the rear is as natural as the multiplication tables. It would be vain to quarrel with it; no headquarters has yet failed in war because of possessing too much information. But it is a little bit absurd when contrasted with the paucity of information along the fire line, where information is like shot and powder, and the word said in time may mean a victory or save a hundred lives.

To reverse the flow, or rather to equalize it, so that all levels may be served according to their necessity—there, truly, is the real problem, complicated to excess by all of the physical stresses which attend the movement of battle forces, but likewise complicated unnecessarily by the blindness and the indifference of men.

It is a truth beyond argument that full and accurate information becomes most vital at the point of impact, for unless it is correctly applied there, the wisest plans of the ablest general will likely fail. But the organization of tactical information during combat runs directly counter to this principle, almost as if it followed an unwritten law—the lower the rank of the commander, the less he is entitled to know about his own affairs.

Much of the time it is the fate of the relatively few men who compose the combat line to move blindly into battle like a colony of moles and to grope for information which the rear is in a position to supply them. They are treated as if the far purpose of combat reconnaissance is to procure information for the higher command, whereas it is crystal clear that the true object in the passing back of all information is to enable the headquarters to further the advance of tactical troops.

Logic would require the total organization by the rear of all information which might prove useful to the front. But the reason why action does not always conform to logic is that many headquarters people become strangers to the front and cannot speak its language or understand its tribulations.

The ever-growing tendency on the part of staff to use the wire or radio for all purposes, thereby avoiding danger and making certain of being at hand when the "old man" wants comfort, is a great block to information. It is a habit which prevents the commander from getting the facts which are most vital to his main purpose and it dissolves the very stuff of co-operation between command and the commanded. Yet how frequently in battle does one observe a higher staff functioning as if its *raison d'être* is that its separate members may take down duplicate telephone messages and then sprint fifty feet to see who can first give the information to the chief! It is bad enough that this is an inexcuseable waste of valuable talent; it is worse that it encourages in the lower ranks a contempt for the character of leadership, for they will excuse waste motion in themselves but they will not pardon idleness in those whom they hold responsible for their own well-being.

Yet in our Army this is by no means a deserved contempt. The average American commander is a man of high conscience. He is never loath to share danger with his men when the sharing of danger becomes part of a necessary function rather than an act. But under the rapidly changing conditions of modern warfare he lives among riddles and is beset by perplexities about how to make best use of his personal force.

The diabolical effect of even such a relatively simple instrument as the field telephone is that it may come to command the commander. It chains him to a system of remote control. At first he sees it only as a useful channel for quick communication in combat. Then he fears to leave it lest it should require his presence in headquarters the moment after he leaves to go forward.

This state of mind in turn creates its own illusion, fostering the conclusion that under the new system of war all matters can as well be settled at a distance, all problems arising within the zone of fire can be fully understood without ever going there, and all moral values which once attended the commander's effort to impress his men with his personality and character are somehow sundered by the new technology of operations. Out of sedentary generalship arises the evil of troops which, while obeying mechanically, have no organic, thinking response to the commander's will.

Yet I do believe that any of the fundamental requirements of fighting forces have been changed by the conditions which modern warfare have imposed on the commander. Not less now than in Frederick the Great's day the essential spirit of the undertaking can be set forth in his words: "The commander should appear friendly to his soldiers, speak to them on the march, visit them while they are cooking, ask them if they are well cared for, and alleviate their needs if they have any."

With his characteristically warm humor, General Eisenhower has commented on the value of the personal factor in the commander's relation to his men under the conditions of modern war. "I found that it did a great deal of good to get down to troops in the combat area," he said. "My presence relaxed them and made them feel more comfortable about the situation. But I was not deceived as to the reason. I knew what was running through their minds. They were saying to themselves, 'There must be less danger than we thought or the old man wouldn't be here.' "

It can be agreed, however, that there is a distinctly new problem created by the complexities of modern war and the distances separating the various levels of command. To every good commander in battle there comes many times this question: How do I reconcile the fact that my post of duty is at the rear with the need that my presence should be felt by my fighting men?

The truly inspired leaders that I have seen in combat or of whose system in World War II I had occasion to make first-hand study—those who did the best job of satisifying the major requirements of over-all control and moral help to the line—invariably made the most of every opportunity to inquire into the physical characteristics of the fire fight. That was a rule of thumb with them—to learn the small details of how the battle was being fought from the men who were carrying the action. They made this a part of their mission any time that they went forward and they invariably adhered strictly to the rule that every trip to the front should be an essential mission, that front-line inspection should have a specific combat purpose, and that their presence should never in any circumstance entail any additional hazard for the combatants.

In this way, they not only used their visits to inspire their men but they learned much which could be applied to conserving the lives and building the efficiency of their forces. These are the two main objects of search along the fighting line. They do not differ radically from the fundamental purposes of sound inspection anywhere else. The difference is that the processes of the battlefield evolve rapidly, so rapidly in fact that no man may keep pace with them through knowledge gained second-hand. The knowledge of the stresses and strains upon human nature, which is as vital to an appreciation of the situation as the counting of man strength and weapon strength, comes only of experience acquired first-hand. There are undertones of hand combat which will always remain unknown to the commander who neither goes himself to the front nor makes it a part of his system to keep members of his staff shuttling between the headquarters and the fire line, and there are degrees of personal fealty which will ever be denied the chief who remains only a name to his men.

The front is to be seen and its conditions are to be understood only through the eyes and words of the men who fight there. There is no interlocutor who can fill in for them. Getting up to a battalion CP and talking to its people about the conditions of the fight will not usually satisfy the object. For while it is true that the competent battalion commander will always know what his men are experiencing, it is not less true that the inefficient commander will not know, though he will rarely plead his own ignorance. One who tries this a few times quickly comes to accept it as a working principle.

The need that a commander be seen by his men in all of the circumstances of war may therefore be considered irreducible. Not to exercise that privilege is to deny his command an additional measure of moral strength which may not be gained in any other way. In fact it may be set forth as a mathematical rule: The values which derive from inspection and personal reconnaissance are in direct ratio to the difficulties of the situation.

Here, once again, there is need for a transfer of thought. In rear areas the commander, high or low, wins the hearts of men primarily through a zealous interest in their general welfare. This is the true basis of his prestige and the qualifying test placed upon his soldierly abilities by those who serve under him. But at the front he commands their respect as it becomes proved to them that he understands their tactical problem and will do all possible to help them solve it.

This is a part of the nature of the combat soldier. He perforce takes a professional interest in the professional problem because his life is at stake in it, though it takes the front to emphasize that fact so clearly that it is reflected in his attitude toward his superiors. When this change occurs, he will stand iron rations and the misery of outdoor living in foul weather for indefinite periods, provided that his tactical experience makes sense and he remains convinced of the general soundness of operations. Once he loses that faith, it becomes very difficult to restore it. He will lose it very quickly when he sees that casualties are wasted on useless operations or when he begins to feel that he is in any respect the victim of bad planning or faulty concepts. Then, he responds to that principle which was once well stated by General James G. Harbord: "Discipline and morale influence the inarticulate vote that is constantly taken by masses of men when the order comes to move forward—a variant of the crowd psychology that inclines it to follow a leader. But the Army does not move forward until the motion has carried. 'Unanimous consent' only follows co-operation between the individual men in ranks."

That is a point for all commanders to remember when they urge men forward in battle. In the relatively peaceful atmosphere of any headquarters more than 1500 yards from the front there is usually a predisposition to calculate the situation in terms of which segment of the line is farthest advanced and to take this as the measure of true progress. Let those who doubt it follow the flow of staff coversation in a representative dozen division war rooms! They will find that such comparisons are frequently made and that commanders not infrequently act upon them without knowing any of the determining circumstances.

It follows that orders oftentimes ignore the nature of battle because they override the prime fact that action, if it is to be decisive, must develop according to the distribution of enemy forces. That does not imply that one's own force must move by the shortest line straight to the heart of the enemy's area of greatest strength. In most cases it will mean the opposite, with the maneuver evolving around the idea of destroying the heart by pinching off the arteries.

But above such elementary concepts it means emphatically that a first responsibility of the tactical commander at every level is to determine, as exactly as possible, by all means within his power, where that heart is located, and then plan his battle or rearrange his plan accordingly.

There are limits to what preliminary reconnaissance can

accomplish. It may often fail altogether. Or it may succeed just enough to convey a false idea of enemy situation.

In either case maneuver against the enemy becomes the prime means of redressing the course and of determining the true situation. All combat is in this sense exploratory. When in the course of operations the true situation is made clear, the commander who thereafter holds rigidly to his original plan, whether because he is too dull to appreciate what has happened or too indifferent to change over, must be regarded as having failed his troops in the most vital particular. It is a failure without excuse.

"An absolute doctrine is impossible," writes Major General J. F. C. Fuller. "For once a doctrine and its articles become a dogma, woe to the army which lies enthralled under its spell."

No truer words than these have been written about infantry fighting, or about the training for it, even when that training is given in immediate anticipation of battle and when the nature of the terrain is well known. Every new battle terrain presents a fresh variety of tactical problems and requires novel adaptations of old methods; moreover, these problems cannot be seen in full proportion until troops have actually arrived on the battleground. That is why it must be accepted as a principle that training carries on into the battle zone until the end of combat and that there is no release from it, even for the best of troops.

Human imagination is not infinite. Given every advantage of accurate intelligence and accurate mapping and air reconnaissance, the most efficient trainers can anticipate only imperfectly how ground not yet encountered will effect the movements of men and the value of weapons. The problems which pop up and provide the most serious harassment are invariably the ones which were not even foreseen. We saw in the Pacific how, having taught infantrymen the rather simple technique of handling explosive charges, we lost men time and again because they lacked an engineer's concept of the limitations of explosives.

Once in discussing with Lieutenant General Walter Bedell Smith certain of the tactical difficulties of the Normandy campaign, I asked whether some of our faults there could be traced to lack of advance information about the bocage country and a consequent pinching of the tactical preparation.

He answered: "Not at all! That wasn't the source of the trouble. The information which we had from the French was more than adequate. Moreover, Field Marshal Sir Alan Brooke and General Sir Frederick Morgan had both come out that way in 1940. They told us about the country, describing it quite accurately. They were very pessimistic about our chances of coping with it. But we couldn't believe what we heard. It was beyond our imagination. The fact was that we had to get into the country and be bruised by it before we could really take a measure of it."

This will ever be the case in war. It is not within the ingenuity of man ever to fully close the gap between training and combat. Once that fact is fully grasped, we have no choice but to incorporate its meaning into the working philosophy of training. So doing, we can arrive at a fresh application of what is really intended by the somewhat vague statement that "plasticity of mind" is the desirable mental attitude in the commander. Thus by a rough approximation: 60 per cent of the art of command is the ability to anticipate; 40 per cent of the art of command is the ability to improvise, to reject the preconceived idea that has been tested and proved wrong in the crucible of operations, and to rule by action instead of acting by rules.

Small unit commanders can well be counseled that training theory seeks the advancement of general principles rather than absolute ideas. The final clarification of tactical ideas

—of methods of attack and defense, and of maintaining a proper balance between the two so that tactical forces may be both conserved and exploited most efficiently—will come only after troops deploy on the battleground. The ground itself is the great teacher; one must be ever ready to apply its lessons with a fresh mind. In every scene thereafter, the lessons are subject to modification not alone according to changes in the countryside but according to changes in the enemy's intent and attitude, and in the prevailing weather.

This last is a topic deserving of far more respectful and comprehensive study than it has ever been given by professional soldiers. There is no more startling omission in the library of war than the failure of all armies and military scholars to deal with it competently. Clausewitz calls attention to the importance of weather and its effect on operations, but he does not define its influence, and he speaks of adverse weather only as a factor which multiplies the frictions of operation.

Surely this cannot be enough to satisfy those who are thirsty for knowledge which will be of practical benefit. It should be said that just as changes in the ground produce new special and general effects on tactics, superinducing changes in the methods of applying fire, so does changing weather impose new tactical conditions on the same ground. The prime effects are not always in the realm of logistics, as Clausewitz suggests. History holds abundant proof that military forces may sometimes triumph by making advantageous tactical use of the same conditions of weather which are imposing a drag on supply. We who have served in the Army of the United States have behind us now a body of experience with weather and with climate in all latitudes such as is unknown to any other nation. We have had the opportunity to learn practically all that there is to know about the effects of weather change upon fighting. There is scarcely a combat soldier who has served in any of the theaters but can recall some vivid lesson from first-hand experience. It remains to be seen whether these lessons can be collected and codified so that in the future there will be no need to learn them all again the hard way.

As an example of the kind of penetrating inquiry which is required, I suggest a comparison between the Ardennes and the defeat of the Fifth British Army in the great Michel Battle around St. Quentin and Peronne, France, in March, 1918. One reading the history of the latter operation cannot escape the impression that the fact of the fog closing around the British defenders was a decisive contribution by nature to the success of Ludendorff's offensive. The playwright stressed it in *Journey's End*. Historians attach almost the same decisive importance to the weather as to the effect of Ludendorff's tactics of infiltration upon an already overstretched British line. Such is the emphasis given this one point in explanation of a great defeat that it might well warrant the conclusion that a main effect of fog is to paralyze the defender.

The soundest reason for rejecting this judgment is that it flies in the face of common sense. What is the usual effect of fog upon the affairs of men? Simply this, that it curtails all movement. The general effect upon military forces is exactly the same as upon a peaceful society. There is no way to change it.

In limited operations, fog or any other conditions of low visibility will enable well-trained troops to traverse an otherwise impassable field of fire, close with the enemy and defeat him, provided that the fog holds and provided that the enemy is of inferior moral quality when dispossessed of his superior fire position. There are many examples of successful small-scale operations under these conditions from the Italian theater. But their significance need not be exaggerated. The more extended the movement, the less is the prospect of success. The hazard of fog falls heaviest on moving bodies. Defending troops have no reason to fear it so long as they remain confident. But it reduces the integrity of attacking forces, hampering communications and multiplying the difficulties of keeping contact.

This one lesson from the Ardennes operation, December, 1944, should be plain for all to read, and we should not be deceived by the primary fact that, having postponed the offensive several times through November because of the unreadiness of troops, the enemy at last struck deliberately from behind a curtain of fog, proposing to turn this to his own advantage. The history of the German staff planning for this operation makes clear what factor weighed most in the decision. Said Colonel General Alfred Jodl: "Mid-December was a very foggy high-pressure period. On December 13 an extended period of fog was predicted, broken about midday. Fog came on December 14. By December 16 we were already afraid that it would lift. But on December 14 Hitler himself had made his decision and had given the order that the attack would begin two days later."

And why was the German High Command afraid that the fog would lift? Again we have the words of Jodl: "Our object was to avoid your air attack. We had hoped to stage the operation in November because that is the worst flying month. Apart from this factor, the foggy weather gave us no advantage, but rather disadvantage." That is what it boils down to—that in the Ardennes fighting the enemy considered that fog would give his ground formations additional protection against a vastly superior Allied Air Force, but he was not unaware that for this first advantage he would have to trade off part of the tactical unity of his assault formations.

The penalty was not made evident so long as the attacking columns were able to retain the road. But it was exacted anywhere that the spearheads were brought to a halt by determined defensive fire and the infantry and tank formations were compelled to deploy over open country. Jodl traces the beginning of misfortune to the "lack of frosty ground which might have supported our tanks" and "the failure of surprise after the second day." But there is more to it than that, though the rest of the story is to be found in our own records.

The complete study of the Ardennes fighting during the period December 16–January 10, in which fog was prevalent except for a few days just prior to Christmas, shows that in most instances it was the attacker who became the victim of local surprise and who most frequently failed because his forces could not keep juncture. This holds true even after the roles of the opposing forces are reversed. The fog did not discriminate. When the tide turned and we began to counterattack—as when the 17th and 101st Airborne Divisions tried to break out of the Bastogne Salient in the early days of Janurary, 1945—the fog did not spare us and we paid heavy toll to the weather which was abetting the enemy guns. As an eyewitness to these engagements, I would assert that the local disarrangement of our offensive operations resulting from this one factor was more considerable than from all other causes put together.

What has been said here about fog is scarcely more than a footnote for a general study of how changing weather influences the form of tactics, though principles remain steadfast. That subject is deserving of far more exhaustive treatment and careful analysis than are within my means. It is sufficient for the purposes of this writing to point out that in battle the scene may change frequently even though the terrain factor remains constant, and that as the scene changes, so may the tactical situation. The contour and accident of ground are not more important to the regulating of fire than are the condition of the earth itself and the state of the atmosphere above it. Torrential rains may vastly reduce the capabilities of automatic weapons. In time of cold and heavy snows the cover given by small villages and wood patches which normally would have little tactical value may become far more prized than the adjacent high ground because it offers temporarily a greater chance for survival. Too, moral values change with every shift in the weather. Rain and heavy skies are as greatly depressing to the will of fighting troops as they are impeding to the mechanics of movement. On the other hand, we learned at Bastogne that fog may become a moral buoy to defending troops once they have learned not to fear it, since it is the one condition most likely to produce the opportunity for clear firing on live targets. Amid the shifting scenes, however, one general proposition holds: The more rapid the change, the greater becomes the influence of reconnaissance and the spreading of information in giving impetus and rhythmic flow to all operations.

During war, it oftentimes happens that one company, by trial and error, finds the true solution for some acute problem which concerns everyone. But when that happens to a company, I can assure you that it is the exceptional company officer who takes the initiative and passes his unique solution along to his superiors even after he has proved in battle that the idea works. A good company idea in tactics is likely to remain confined to one company indefinitely, even though it would be of benefit to the whole military establishment. Such omissions are not due usually to excess modesty or indifference on the part of the officer, but to his unawareness that others are having the same trouble as himself. Combat is a business in which every man's horizon is greatly foreshortened. The commander's load of responsibility is such that he inclines to act negatively toward even the other units in his own echelon. He thinks concentrically. He spends so much time thinking about how the other elements can serve him that it becomes difficult for him to reverse himself and give some portion of his thought to how he may serve them. Thereby much is lost from unity.

It is a state of mind, however, which is not wholly beyond remedy, provided that the officer is caught in time. The besetting fault in our hasty wartime system of schooling young men for the leading of combat troops was an exaggerated emphasis upon the "book" or approved solution. Though the importance given the book in our military thinking is not without its merits, it is not conducive to the development of an officer corps which dares to think original thoughts. Constant change is the natural order of war. New situations frequently call for hitherto undreamed solutions. The maximum efficiency therefore comes of a system of thought which assures that each new and sound idea will be circulated promptly.

As I pointed out in the beginning, the fundamental purpose of all training today should be to develop the natural faculties and to stimulate the brain of the soldier rather than to treat him as a cog which has to be fitted into a great machine. The true purpose of all rules covering the conduct of warfare and all regulations pertaining to the conduct of its individuals is to bring about order in the fighting machine rather than to strangle the mind of the man who reads them. We have taken some promising strides in this direction, such as the modern tendency of the service schools to eschew the "approved solution" and to offer instead several possible solutions. That gives some play to the imagination and it is imagination primarily which distinguishes the brilliant tactician from his plodding brother. Sherman said of Grant at Vicksburg that he—Sherman—could have carried off the campaign more effectively, but that, unlike Grant, he could not have dared to think of it in the first place.

What we need to aim for is greater freedom of professional thought by all ranks and a philosophy of command which is consistent with this general purpose. If we can achieve that, we will have prideful and satisfied soldiers at all levels and we will have a satisfactory Army. I am not overlooking the inevitable percentage of malcontents. With respect to them it is sufficient to say that while we will ever have them with us, their opinions should never be permitted to influence policy even though they speak with the voice of Stentor. Policy can remain strong only so long as it faithfully serves the best interests of the majority of dutiful soldiers, officers and men alike.

Loyalty in the masses of men waxes strong in the degree that they are made to believe that real importance is attached to their work and to their ability to think about their work. It weakens at every point where they consider that there is a negative respect for their intelligence. This rule applies whether a man is engaged in digging a ditch or in working up a loading table for an invasion. What he thinks about his work will depend in large measure upon the attitude of his superiors. The fundamental cause of the breakdown of morale and discipline within the Army usually comes of this, that a commander or his subordinates transgresses by treating men as if they were children or serfs instead of showing respect for their adulthood. But I submit that it is not possible for an officer corps to keep its aim high in this particular if its own members have been consistently fed with the spoon. That has been the source of some of our difficulty.

The requirements of modern war are such that we certainly do not want to turn out one soldier identically like another. But the rule applies to officers as well as men. The greater freedom which is needed has nothing to do with social behavior or privilege. It is the freedom to think boldly for the common good, for, as Kant has said: "What one learns the most fixedly and remembers the best is what one learns more or less by oneself."

To square training with the reality of war it becomes a necessary part of the young officer's mental equipment for training to instill in him the full realization that in combat many things can and will go wrong without it being anyone's fault in particular.

War is aimed at destruction. The fire and the general purpose of the enemy are directed against one's own personnel, matériel, and communications, with the object of keeping one's own design from coming into play. Small plans miscarry because the wrong man happens to be hit at the critical moment or the guns which were counted on are knocked out of action.

The problem of command in battle is ever to establish a safe margin which will allow for such misadventure. But this much is certain—that there is no system of safeguards known to man which can fully eliminate the consequences of accident and mischance in battle. Hence the only final protection is the resiliency and courage of the commander and his subordinates. It therefore follows that the far object of a training system is to prepare the combat officer mentally so that he can cope with the unusual and the unexpected as if it were the altogether normal and give him poise in a situation where all else is in disequilibrium. But how to do it? I would say that the beginning lies in a system of schooling which puts the emphasis on teaching soldiers *how* to think rather than *what* to think even though such a revolutionary idea would put the army somewhat ahead of our civilian education.

In operation Market, the airborne invasion of Holland, a young West Pointer, Lieutenant Colonel H. W. O. Kinnard, had the mission of defending that portion of the corridor lying to the west of the town of Veghel. He was in command of First Battalion, 501st Parachute Infantry. The sector was under heavy enemy pressure from the hour of the drop. Against the advice of his superior he advanced and put into execution a plan for defending the corridor by carrying the battle to the enemy. During a three-day march he made a complete 360-degree wheel through enemy country, destroying enemy forces equal to three times the strength of his own battalion. The alarm caused by this advance reverberated to the highest levels. The enemy alerted a reserve corps to meet the danger as it was believed that the maneuver indicated that the main Allied attack was changing direction.

Here is an example of a soldier thoroughly in command of his own situation rather than permitting his situation to command him. He weighed the hazard that he would be moving at all times with at least one flank exposed, then accepted this risk in view of the prospect for proportionate reward. I know of no better illustration in the book of war of the quality of mind needed in the combat officer.

The test of fitness to command is the ability to think clearly in the face of unexpected contingency or opportunity. Improvisation is of the essence of initiative in all combat just as initiative is the outward showing of the power of decision.

It is a necessary cushion to give all young combat officers a profound understanding of this fact, for it is the first step toward endowing them with the kind of self-confidence which will make them fully communicative with their superiors when they engage the enemy.

An army in which juniors are methodically "covering up" for fear they will reap criticism for using unorthodox methods in the face of unexpected contingencies is an army which is slow to learn from its own mistakes.

An army in which juniors are eager because they have found it easy to talk to their superiors will always generate a two-way informational current.

Such an army will in time develop senior commanders who will make it their practice to get down to troops in quest of all information which may be used for the common good.

This suggests one more point which possibly should be discussed under another heading though I find it convenient to place it here.

When troops have been hard used in battle, and especially when green troops have taken heavy losses during their first engagement, talk itself is the easiest and most effective first step toward the re-establishing of a fighting morale.

Nothing is more likely to break the nerve of an intelligent and sensitive young commander in the aftermath of a costly and bloodletting experience than to leave him alone with his thoughts. That holds true also of the men under him. Men need to talk it out. The need of such a release is greatest when they feel that they have been whipped.

It makes little difference how clearly the circumstances say that the fault was not one's own. The shock which comes of seeing one's own men or comrades killed and of pondering one's own hand in the making of their fate leads almost inevitably to a mood of self-accusation and bitterness—the tokens of moral defeat. The more able and conscientious the commander, the more likely it becomes that he will react in this way. It is only the bloody fool who remains wholly insensitive to his own losses, and he is both bloody and a fool because his very attitude proscribes self-criticism and leads to unnecessary expenditures of force. The ranks are never hardened by death in their midst. Losses are never a help, and unless they are incidental to action which has a clear and practical military purpose, they are weakening to the confidence of troops. This truth is so obvious that it would not be considered worth stating if I had not at times encountered field commanders in our Army who welcomed battle losses and tried to arrange them out of a mistaken belief that it helped harden the other men.

The darkest hour for the novice in war comes with the recoil after the unit has been badly hit. It is then that the young commander has greatest need of the friendship and steadiness of his superior or of any other officer whose judgment he respects. Criticism or tactical counsel are of no value at that time. They can be given later if necessary, but in the wrong hour they add to the hurt. Let him get out his crying towel! When he has told how it happened, the important thing is that he be given a pat on the back, an assurance that he did his full duty, and some little reminder that while he may feel that his losses are excessive, such incidents are an unavoidable feature of combat and do not keep one from coming back in the next round. In Siegfried Sassoon's *Memoirs of an Infantryman,* the young lieutenant tells of emerging from a bloody trench raid and meeting his colonel: "This was a Kinjack I'd never met before, and it was the first time I had ever shared any human equality with him. He spoke kindly to me in his rough way, and in so doing, he made me very thankful that I had done what I could."

If a senior commander cannot do that much in good conscience for the benefit of a junior, it can only be because the latter's personal action was so thoroughly discreditable that he deserves to be relieved on the spot.

The treatment is simple. Its psychological result should be apparent. I have seen it work many times in battle. Major General Archibald V. Arnold was a master hand at this art of cushioning the shock for his younger subordinates. But it is by no means a rule of action even among those senior officers who make a sincere effort to live close to their troops. More times than not, I have seen them freeze up in the very hour when their subordinates were hurting worst for a few words of understanding. Then someone else had to come along and do the job for them. Likewise, I have witnessed units which had been badly bruised their first time in battle, spend several days in painful brooding, and then bound back almost at once when given a little intelligent moral treatment by a superior. A man remains a man after he puts on a soldier suit. Death in the company is like death in the family. Talk relieves tension. It is the first step toward moving on to life's next problem.

However, there is a more positive side to this matter. Success is not easily observed or understood in battle. That is not alone because of the confusion of the scene whereby, in the words of Clausewitz, "The impression of the senses is stronger than the force of the ideas resulting from methodical reflection." More truly, it is because success is measured in terms of total effect upon the enemy, and in the nature of close engagement this is a matter which usually may not be judged at short range.

The unit which fights a successful action but is without knowledge of its success may even insure a great victory for some larger body and still emerge from the battlefield with a feeling of inferiority. It may come as a shock to many soldiers who have not engaged in close study of the processes of the battlefield to be told that this can happen. Yet from my own experience with many companies after combat I would say that the chance is about even that the company which has distinguished itself in a particular action will continue on, not knowing whether it has done well or was a comparative failure. That is just as true of the battalion.

On D Day, at Omaha Beachhead, Company M, 116th Infantry Regiment, did perhaps the most efficient job of fighting, its circumstances considered, of any infantry company in the landing. It was a weapons company and it landed in boat sections in one of the later waves, the presumption being that its part of the beach would already be under control. But such was not the case. They were the first troops to hit this particular sector of the beach. The officers and non-coms determined at that moment that the initial handicap could be overcome only by an initial showing of extreme fortitude. Under intense enemy fire, working patiently with their men, they persevered until every piece of equipment was dragged across the beach. It was the only company along Omaha Beach which didn't leave one piece of equipment or one sound rifleman behind on the sands. With the officers leading and the men following, the Company then advanced up the steep escarpment, in the face of sharp fire from enemy small arms. All day long it moved as a rifle company, hitting hard and traveling far. It stood at the point of one of the thin salients forming the tenuous outline of the beachhead when night closed. That *was* success. But the survivors of Company M fought on through the remainder of the campaign, and through St. Lo and the siege of Brest, still not knowing that on D Day they had done anything exceptional.

If this incident were unique it would scarce be worth mention. But it is typical. Battle is a fog for the men who fight. The small unit will usually remain in the dark about its own achievement unless someone from higher up clarifies "the picture." And that is done automatically in but few cases. The higher level is too engrossed with other problems or its means of research are so inadequate that it does not always know why success has come. Thus it happens that a company or battalion may win a victory under circumstances which make it appear almost as a defeat on the local ground, either because it is overconscious of its own hard losses or because the over-all tactical effect could be seen only at the higher headquarters. This impression of failure will continue so long as no one concerns himself with setting the facts aright.

Thereby is wasted a precious moral opportunity, which in the long run is as wasteful of fighting strength as an outright loss of men.

In combat nothing succeeds like success. The knowledge of victory is the beginning of a conviction of superiority. Just as truly, the savor of one small triumph will wholly drive out the bitter taste of any number of demoralizing defeats.

9

TACTICAL COHESION

"Sing him a new song; play skillfully with a loud noise."

—From THE PSALMS.

IT IS a matter commonly noted that they who write of war tend to use loosely the expression "battle-seasoned troops" as if there were a kind of mental toughening which comes from experience under fire.

The idea is wholly misleading; it mistakes the shadow for the substance. One of the effects of the shock of engagement is that it shakes the weakest files out of the organization. But as for the veterans who remain, they do not grow more callous to danger as they meet it increasingly nor do they ever become more eager for the contest. As they grow in knowledge, the nerve may become steadier in that they are less susceptible to wild imaginings. But if it were certain that battle experience of itself provided the sure safeguard against this evil, then it could not happen that high commanders sometimes fail in crisis because of it.

Take one man! Forget for the moment that he is simply an individual and consider him as a tactical unit. Order that unit into an area of great danger. Attempt to condition it under the varying types and conditions of hostile combat fire.

Then grant it a respite for a time and order it into battle again.

Though this process is continued indefinitely, it will never condition the one-man unit to face fire more resolutely and more intelligently than in the initial experience. Indeed, the effect would always be a steady downgrading and the deterioration of the unit's mental and moral fiber.

What then is the prime difference in the effect on one man and on numbers? It rests in this, that what we call "seasoning" in troops is largely a matter of learning to do the thing well instead of doing it badly.

Since troops do not conquer the fear of death and wounds, it is idle to think of any such basis for the establishment of a combat discipline. The latter is simply the reflection of the growth of unit confidence which comes of increased awareness and utilization of one's own resources under conditions which at first seem extraordinary but gradually become familiar.

Until that kind of confidence is born, there can be no effective action. Green troops are more likely to flee the field than others only because they have not learned to think and act together. Individually, they may be as brave and willing then as during any subsequent period, but individual bravery and willingness will not stand against organized shock.

With the growth of experience troops learn to apply the lessons of contact and communicating, and out of these things comes the tactical cohesion which enables a group of individuals to make the most of their united strength and stand steady in the face of sudden emergency.

Let us take the simplest of illustrations. We found this same example useful during Central Pacific operations in emphasizing to troops what was lost by not keeping juncture. Four infantrymen, each with a weapon, have deployed in face of the enemy and have independently taken positions in four separate shellholes. They are green at fighting and none attempts to call out or to reconnoiter in any direction in search of strength which may help him if the enemy presses. The position is then something like this:

There are four men present on the ground and four or more weapons. But regardless, the position has the effective moral strength of only one man. In the face of any sudden alarm its tactical strength is unlikely to be any greater than its moral strength. So long as the four men remain unknowledgeable of each other, that is the prevailing condition. Each man is likely to give ground because he considers himself unsupported. It is more than probable that the total ground would be yielded if one determined rifleman came against it, for the natural assumption of any one of the defenders would be that behind this rifleman were coming others.

But now let us say that one man calls out. As a result of his calling, all four become aware of the presence and the weaponing of their comrades. The moral and tactical change in the situation is nearly absolute. It is no longer a chance scattering of four riflemen over untenable ground. It has become in fact a military position, with all-around security and full tactical advantage of the multiplication of weapons.

Where formerly each man was three-quarters vulnerable in tactical fact, he is now protected on flanks and rear. From the physical uniting of the position comes its moral solidarity. The means establishes the will.

More than that, it is altogether probable that in the course of getting together and considering what should be done to cover and protect each other, they have related what they are doing to its effect on the larger unit of which they are members. While three of the men hold, one crawls rearward or flankward to locate the nearest elements of the company and *to make certain* that these elements notify the commander of what has happened in the forward ground.

The unit of four is now supported. It has a line of communications. It has provided full information. It has, in fact, satisfied all of the requirements of sound tactical procedure. In consequence, the position now draws strength from the whole command.

I feel sure that at this point I hear a protest from my reader that I am talking about the ABC's of minor tactics and that nothing has been said that is not already well known.

That much is granted. But now the question: Is it fully appreciated that the most general cause of small failures along our combat line, which frequently promote the confusions of larger bodies of troops, is the individual failure of the American soldier to respect this simple but fundamental principle? Our aggregate tactical weakness stems largely from this failing. We have encouraged the man to think creatively as a person without stimulating him to act and speak at all times as a member of a team. The emphasis should be kept eternally on the main point: *His first duty is to join his force to others!* Squad unity comes to full co-operation between each man and his neighbor. There is no battle strength within the company or regiment except as it derives from this basic element within the smallest component.

As to the four men mentioned earlier, the odds would be very much against them taking the spontaneous action which would bring off the union of forces. A chief fault in our men is that they do not talk. They are not communicative. In combat they are almost tongue-tied.

In Europe they were frequently astonished at the incessant talking and shouting that went on among the enemy formations during an action. They mistook it for naïveté in the Japanese that in combat they frequently acted in the same way. That there was a direct connection between these methods and the phenomenal vigor with which our enemies organized and pressed their local counter-attacks seems scarcely to have occurred to our side.

But along our own fronts, contact was frequently broken and many small actions were lost because our men had not learned that speech is as vital a part of combat as is fire. Under the pressure of battle they could not act consistently according to the principle that the full distribution of pertinent information contributes as greatly to defensive strength and to offensive potential as does the proper disposal of weapon power.

Now to define pertinent information in its simplest terms: It is any information which will enable the uniting of strength, which in turn means the multiplication of strength.

It starts with the smallest tactical unit—one man—what he knows of the situation and intentions of the men who are next him and what he can tell these men of his own situation and intentions.

I think I can make it clear further along in this discourse that in the individual as in the unit, the search for information and the giving of it are the true beginnings of what is called initiative. For the present it suffices to say that the greatest possible stimulant to the initiative of the commander is for the subordinate to continue to supply him with all information as it develops, and that conversely, to build initiative in his men, the commander must keep them informed of the general situation, the object, and the role of all elements.

But we cannot drop the subject at that point. The rule applies equally between soldiers who are in a position to act together as between the commander and the subordinate. If each file, or every squad and platoon leader, waits for the commander to inform him what is happening on his immediate right and left, then he will never hear and he will never unite his action with theirs. No company commander can keep pipelines to that many men; the commander who knows his business won't even attempt it.

What was said earlier in the chapter about the four men and the manner in which their strength matures applies as directly to multiples of that number. Strength will multiply and decisive action will become possible at the rate that information flows to all concerned.

The total strength of the command is the total of what all ranks know at the given moment about the strength of the command and of the position.

While I was in the Central Pacific, I was given the problem of determining why it happened that when fire came against an advancing infantry line, the resultant check to the line would invariably cause a delay of forty-five to sixty minutes. The phenomenon had interested Major General Archibald Arnold and he wanted to know the basic cause. The troops were "battle seasoned." The tactical terrain was not abnormally difficult. The flat, sandy islands were fairly clean. Though they were thickly grown with palm, mangrove, and pandanus trees and tropical shrubs, the bush was nowhere impenetrable. So the problem did not appear to be insurmountable.

We researched this question through eleven infantry companies and one reconnaissance troop, all just out of combat. What we found is best related in the language of the official report:

> When an advancing infantry line suddenly encounters enemy fire and the men go to ground under circumstances where they cannot see one another, the moral disintegration of that line is for the moment complete. All organizational unity vanishes temporarily. What has been a force becomes a scattering of individuals. This is inevitably the case. Men going forward in line are in sight of one another. They therefore have a sense of unity. But when they come unexpectedly in check and go to ground, they no longer have knowledge of the position of the men on left and right because they have not taken true cognizance of these things as they moved along. Indeed, it is not possible for them to do so if they are to be alert to the danger which lies ahead. While erect, they feel the presence of the others; when they go down, this feeling is lost. The platoon leader no longer feels the whereabouts of his squads; the squad leader is certain only of his own location.
>
> Before the company or group can again become a going concern, capable of working its will on the enemy, it must reintegrate, and before it can reintegrate, communication must be restored between the fractions. This does not

necessarily mean word communication. It may be only the communication which comes of seeing that others are present, though before this can well happen, there must be some precipitating act.

It may be done through one bold individual standing erect and saying to a few others: "Follow me! We're going on." If a few arise and follow, the entire line is apt to get in motion. On the other hand, if this same individual advances alone but says nothing, it is unlikely that he will have any followers. One word, "Come!" makes his action tenfold as effective as if he plunges gallantly ahead in silence. The act of moving is initiative; the act of coupling motion with speech is thinking initiative.

Yet it should be observed that reintegration rarely takes place in this manner. The restoration of unity and of impulse to the attack is therefore subject to the inertia and the trial-and-error methods of small unit leaders, who, while remaining down, try to resume contact with their scattered fractions, or else fail even that simple duty. In this way much valuable time is lost. Even when the enemy fire is so misdirected that there is no real physical danger to the advance, once men have gone to ground, a minimum of ten to twenty minutes will be lost before small unit leaders take the initial steps to re-establish control.

This lag occurs because there is no standing operating procedure covering such a development. The small unit leaders have not been given a direction during training to some such effect as this: "The first thing you do as you hit the ground is re-establish contact with your men. Determine where they are and let them know where you are." The problem has never been surveyed as a training matter and hence such partial solutions as are to be had on the field of battle are not of general application even within the company.

In fact, the majority of small unit leaders do not take any steps toward restoring control, from which alone can come unity of action. Some try to contact their men by voice or by relay of voice; during an action, while the men are prone, the voice will rarely carry more than twenty-five feet; this means that unless there is relay and all hands understand what is being attempted, the voice method is ineffective. Others try this plan: They look for higher ground immediately ahead which is still under cover, then crawl up to it so their own men will see them. Still others wait for a tank to come along, then use it for cover as they walk across the front.

That there is such a variation of method tends only to emphasize the conclusion that there can be no solution of the problem until there is full consideration of it as a factor which seriously deprives the infantry attack of its momentum.

Once halted, even if there has been no damage, the line never moves as strongly or as willingly again. After three or four such fruitless delays, men become morally spent rather than physically rested. All impetus is lost and the attack might better be called off for the day.

The report has been quoted at considerable length because, while it treats of a special tactical problem, the problem itself is symptomatic of one of our general difficulties in the maneuver of troops in battle. With slight variations one found the same problem in Normandy, Holland, the Ardennes, and right up to the last hour when we crossed the Elbe. Within the small unit the loss or lack of control was almost invariably traceable to the failure of close-up communications. In most cases this failure was not caused by anything the enemy had done but by the neglect by our own soldiers of common-sense measures which were well within their means.

The physical characteristics of the Central Pacific atolls were such that the disintegration of the unit (the splitting of the body into unconnected fractions) happened very quickly whereas on the fields of Europe the process was more gradual. But the net results were about the same in both Theaters.

Nor is it too broad an assumption that this is a most normal aspect of infantry combat in our times. The company, coming under fire, literally begins its engagement by falling apart. Thereafter, so long as it continues to engage, the overriding problem of the commander is the reunification of his elements.

Proper fire support and direction are among the tools which he uses in bringing about cohesion. But the fundamental means is communication—getting his men to link up by talking to one another and then sending along word of what they are doing and what they have seen.

Now this is a very simple thing. It is so simple that it recalls the warning from Colonel G. F. R. Henderson: "In war the simple things are the most difficult."

Thus far, however, I have considered only a few of the results of the difficulty and have said nothing of its causes. Once again, I feel that it stems from a definite blind spot in our training. Let us see how and why.

The proper aim of training is to overcome the inhibiting effect of fire and of danger upon the individual and by so doing bring about unity of action. I hold this truth to be self-evident, since it can scarcely be denied that brilliant individualism at the expense of team play will invariably prove more fatal on the field of battle than ever on the field of sport.

Therefore I think we can put it down as an axiom that initiative is a desirable characteristic in a soldier only when its effect is concentric rather than eccentric: the rifleman who plunges ahead and seizes a point of high ground which common sense says cannot be held can bring greater jeopardy to a company than any mere malingerer.

But then, how do we proceed toward the development of a concentric initiative in the individual? It is in seeking the answer to this key question that we discover the significant disparity between the American doctrine of what training should seek to inculcate in the soldier and the techniques which are employed in training supposedly in support of that objective.

We say that we seek initiative in our men, that it is the American way of fighting. We say that we want men who can think and act. We are just as steadfast, however, in proclaiming that the supreme object in training is to produce unity of action. These two aims are not mutually exclusive; in fact, they are the complementary halves of an enlightened battle discipline.

But the very curious part of it is that training largely ignores the sole principle which makes these two basic ideas fully and finally reconcilable. We do not teach our men from the day they first put on uniform that speech in combat is as vital as fire in combat. We do not say to them that for a man to be able to think straight about his tactical situation is not enough; he must communicate his thoughts to others before they can begin to produce unity of action. Out of speech or from the written word which is its substitute comes all unification of strength on the field of battle, and from the latter comes decisive action. This applies to two men serving together on an outpost; it applies equally to the battalion or the regiment.

Yet it is almost an anachronism that one must even refer to this subject. We are so certain that the speech principle is well recognized that much more thought and time are devoted by our Army schools to the study of how speech can be refined—how messages can be made more brief, ideas more simply stated, and orders more rapidly transmitted—than to examining the means by which men can be made articulate in line of duty. It seems to be taken almost for granted by the trainers that the impulse to send a message or to impart all useful information is so automatic in the normally intelligent combat soldier that it requires almost no special cultivation.

Nothing could be further from the fact! The tendency is ever to smother information in combat, to keep what one knows to one's self, to dismiss the idea that it will have any value to a comrade or to higher authority, to argue that what might be gained would not justify the effort, to conclude that the special facts must already be known to all concerned, or if none of these things, then simply to fail to recognize information at its face value.

Take it or leave it, this is the raw human nature of our average man in battle and we suffer more small tactical reverses because of it than from all other causes put together. It is for this reason mainly that cohesion lags and support fails and commanders soon find their elements getting beyond control. It slashes the fire ratio like a knife, the ratio discussed earlier in the book and resolved in the statement that not more than one man in four ever fires his weapon. Yet a general correction is within the means of the trainers once they understand the source of the difficulty.

Before they can come to that, however, it is first necessary to know what are the roots of initiative in combat. The ample literature covering the subject of training troops for war is cluttered with high-sounding but vague definitions of initiative. The words of Colonel F. Gory serve as an example: "Initiative in the soldier is the quality in virtue of which he decides to act on his own as soon as his immediate action becomes useful," or the equally sterile definition by Commandant Leroux: "The initiative consists in freedom of the choice of means which can be employed to attain the prescribed end." One can quote such beautiful words until hell freezes over. But for all the practical good they do they might as well be allowed to escape down a barracks rainspout.

The question still remains: What kind of initiative is beneficial and what kind is harmful and how may troops be taught to distinguish between the two? Until a clean line is drawn it is a questionable practice to lecture troops on the value of initiative or to give them the idea that it is the *sine qua non* of the good combat soldier.

I believe that it is possible and profitable to chart a relatively clear course on the matter. Already, I have emphasized that the act of willingly firing against the enemy, under the conditions of the modern fire fight, should be regarded as an instance of high initiative in the infantry soldier. Also, earlier in this chapter, I made mention of the "thinking initiative."

This is its meaning: It is the soldier acting on his own to advise others of his tactical situation or conveying any other information which may be of general benefit in furthering the tactical situation of the company or in enlisting the aid of others in carrying out any action which will benefit the tactical situation of the company.

It is my belief that training cannot safely go beyond this definition in stimulating initiative within the ranks; and I feel further that once we have evolved a system of training which satisfies this theory of the initiative, we will be well on our way toward the attainment of the highest possible standard of combat discipline. An army in which all ranks are indoctrinated from the beginning with the knowledge that fire and person-to-person communication are the twin essentials of successful minor tactics will generate spontaneity of action and reuniting of effort in the face of any battlefield emergency. When the two ideas become linked and the principle dominates the thoughts of fighting men, they will have a new kind of confidence in the face of the enemy and that confidence will further the conservation of force at the same time that it stimulates aggressive action. But so long as these things remain unrecognized, no army will fully exploit its fighting potential. It will sell short the spirit of the fighting soldier even though it strains to the limit to satisfy his bodily needs and to keep him happy.

In the Army of the United States we act toward speech as if we were mortally afraid of it. We tell our men to think; yet we never tell them that if in combat they remain dumb, it is slow suicide. Far from exploring new ground, we have even lost some which we once held; I can vouch from my own experience that the combat infantryman of World War I was far more given to "talking it up" during battle and to shouting directions to his fellows and discussing the situation with anyone close around him than the man from World War II.

This last one was about the mutest army that we ever sent to war. Nor is that to be wondered at. The period between wars was an age of rapid advancement in communications technique. Radio was born; the telephone was vastly improved; the teleprinter appeared and television waited just around the corner. It all came so fast that we were struck dumb by our own magic. Impressed by the completeness and efficiency of our channels, we tended to forget that no mechanical means of communicating ever given man can become a substitute for the spoken word, and none can amplify thought. The origin of an idea is still where it was in the time of Columbus or of Adam.

There was such concern that all of the new techniques be mastered by the greatest possible number of individuals and that invention fully exploit all possible means of communicating that there was all too little time available for re-examination of the basic principle. It lay there, almost lost to sight under a debris of antennae, microphones, and receiving apparatus as we got our men ready for combat.

During basic training they were given no theory of communications as an active person-to-person principle which they would have to apply continuously in combat if they were to survive. Later on, such instruction as was given them on communications dwelt upon the use of the walkie-talkie, the discipline of a message center, the writing of messages, the simplifying of oral instructions, or some other technical subject.

But no one ever said to them: "When you prepare to fight, you must prepare to talk. You must learn that speech will help you save your situation. You must be alert at all times to let others know what is happening to you. You must use your brain and your voice any time that any word of yours will help you or others. You are a tactical unit and you must think of yourself that way. Don't try to win a war or capture a hill all by yourself. Your action alone means nothing, or at best, very little. It is when you talk to others and they join with you that your action becomes important."

No, no one ever said these things. In thirty years of rather close experience in or with the Army of the United States I have never heard one lecture or day-room discussion on the tactical values of person-to-person communication on the battlefield. Moreover, I have never heard a commander refer to the subject.

But I have seen some commanders fail in combat because they had not talked to their men about these things. When the company engaged, though the men fought willingly, they fought to little avail because the words which might have held them together were left unsaid.

10

WHY MEN FIGHT

"General Meade was an officer of great merit with drawbacks to his usefulness which were beyond his control. . . . He made it unpleasant at times, even in battle, for those around him to approach him with information."

—U. S. GRANT in his *Memoirs*.

SO FAR, we have considered speech in combat mainly as lubricant to all of the cogs in the complex mechanism of tactics. As Disraeli said, men govern by words. It is by virtue of the spoken word rather than by sight or any other medium that men in combat gather courage from the knowledge that they are being supported by others. Battle morale comes from unity more than from all else and it will rise or fall in the measure that unity is felt by the ranks.

However, the tactical effect of speech is not alone that it furthers cohesion, from which comes unity of action, but that it is the vital spark in all maneuver. Speech galvanizes the desire to work together. It is the beginning of the urge to get something done. Until there is speech, each soldier is apt to think of his situation in purely negative terms. With the coming of speech he begins to face up to it. Let those who doubt it place themselves among several men who have just been pinned by sniper fire at close range. What happens?

These men will hold to earth or get in close to fallen timber; but they will do nothing constructive about their situation until one of them makes a concrete suggestion: "It's too hot; let's get out," or "You cover me while I work up to that tree line."

In movement during combat the greater danger to the commander is not that he will err in wording his order but that he will not follow through in making certain that the order is heard and understood all along the line. Words repeated out loud down to the last man will be obeyed. But an order only half heard becomes a convenient excuse for noncompliance. In the giving and in the relaying of orders the rule is to remove every element of doubt. If there is not time for this precaution, there is not time for the maneuver.

At any stage in battle, whether troops are attacking or defending, warmth in the giving of an order is more to be desired than studied self-containment. Too much has been said in praise of the calm demeanor as an asset in a fighting commander. That may have its place at the higher level. At the lower levels men do not fight calmly and they are not reassured by commanders affecting the manner of an undertaker or the dead pan of a poker player.

One of the finest river crossings in our army during World War II was the establishing of the bridgehead across the Elbe River near the city of Magdeburg by the 331st Infantry Regiment on April 13, 1945. Another American unit had held a bridgehead just above this point but had been beaten back by an armored counterattack and forced to retire to the west bank of the Elbe. Nonetheless when the 331st Regiment began its attack, Colonel George B. Crabill had his lead battalion in small boats moving across the open water within less than thirty minutes of arriving at the river bank. They were set on the far side of the Elbe before the enemy knew that a fresh advance was under way. Not one rifle had been fired.

The attack had begun soon after midday. At nightfall a platoon of tanks and the body of the Regiment were brought across the Elbe. The enemy began a strong counterattack shortly afterward with infantry, artillery, and armor. This pressure continued for three days but the defensive position held solid. The fighting had died by the time the first contact was made with the Red Army farther south along the Elbe. Asked how he had been able to move so fast in the first place, Crabill replied: "Because I was more excited and enthusiastic than my men. When we got to the water's edge, I moved along the line of my men, giving them a love kick in the butt. I kept shouting to them as I moved along: 'Don't waste the opportunity of a lifetime. You're on the way to Berlin. We can get there. You can cross now without a shot being fired. But you got to move *now*. Don't wait to organize. Get into those boats! Get going!' " That's the kind of stuff I mean. It's the touch that men understand.

When a retrograde movement becomes necessary in combat, it is an invitation to disaster to move before men are told why they are moving. If the pressure has made that fact obvious, then they still must be told how far they are to go and the line or point to which they are withdrawing must be made clear and unmistakable. Otherwise they will keep moving and all control will be lost. The spoken word is the greatest of steadying forces in any time of crisis. An excited lieutenant shouting: "Get the hell out of here and follow me to that tree line on the far side of the creek," will succeed, though a perfectly calm captain, trying to bring off the same movement but keeping his voice down with the result that the men do not hear him will fail. Formal language under these circumstances is almost unknown in the Army of the United States. In fact, "Get the hell out of here!" has virtually established itself in our jargon as the customary order.

There is a further point while on the subject of a fall-back.

Wherever troops engage, they should be told on what ground they are fighting, if this is possible. It need not be done with co-ordinates, which in any case are hard to remember. Some such phrase as "the hill north of St. Mary's farm" or "east of the village of Grand Pre" is sufficient. This is a safety measure in the general interest, and to ignore it is to waste a type of insurance which costs very little extra effort. In the event of a general reverse the worst form of operational confusion ensues when the survivors come drifting back and are unable to give an account of where they have been. In large-scale operation the effect is paralyzing. It means that command is denied the most vital intelligence of the movements of the enemy. It opens the door to him so that he can pile surprise upon surprise.

We saw this happen in the Ardennes fighting in December, 1944. Not one in a thousand of the stragglers falling back through the American lines could say where they had last engaged the Germans. This was not primarily because they were suffering from shock but because they had never been told by their leaders. Yet many of them had been routed from ground which the unit had been holding for days or hours. The commanders knew what they were defending but they did not think it was important to tell the men. Therein they were wrong. When all else was obscure, just a little knowledge in the ranks would have been priceless to the higher commands.

Man is a gregarious animal. He wants company. In his hour of greatest danger his herd instinct drives him toward his fellows. It is a source of comfort to him to be close to another man; it makes danger more endurable, like hugging a two-inch sapling while sitting out an artillery barrage. Since this is his natural urge, what restrains him and enables him finally to retain his position in the formation which is needed for successful maneuver?

Primarily, it is his training, his intelligence, and his habit working against his instinct. Said General Dragomiroff: "A strong moral education is the best safeguard to the solidarity of troops under fire." Even so, the soldier will forget or discount much that training has taught him as the danger mounts and fear takes hold. It is then that the voice of the leader must cut through fear to remind him of what is required. The reasoned explanation of why this is true has never been more clearly stated than by Staff Sergeant Pete F. Deine of the 17th Infantry Regiment after he had won the Silver Star for taking over the leading of a demoralized platoon during the Burton Island fight:

> I knew that the men were afraid and careless at the same time. Though some were being killed, the others would not take even average precautions in going after enemy installations as we passed through them. I knew they were afraid because I was aware of my own fear. Then I asked myself why it was that we felt fear in each other and I realized it was because all of the leaders had quit talking. I knew then that the only way to get confidence back into the platoon was to talk it up, as a man might do in a football game. I continued my own attack on the enemy shelters and spider holes, but there was this difference, that I now began yelling to the others, "Watch me! This is what you're supposed to do. Get at it. Keep working. Keep your eyes open." Soon the platoon became collected and began to operate methodically. But I kept talking until the end because I had learned something new. Leaders must talk if they are to lead. Action is not enough. A silent example will never rally men.

In battle, the voice of the leader is always needed to call men back from carelessness. It is their chronic attitude in and out of danger. Even in veteran troops it is not the expected presence of the enemy which keeps them alert on a hostile field but the force which they feel pressing them at the given moment. When fire comes against them they sense danger from every direction; unless they are informed of the source of the danger, there is apt to occur a swift moral transition in which they become "mentally pinned" by the mere incidence of fire.

Once the fire lifts, the complete change-over from this attitude is not less marked. Should the same troops then advance rapidly into enemy ground from which they receive no fire, they will quickly abandon all security measures and become lax in all respects unless they are constantly cautioned and cajoled by their leaders.

In an earlier chapter I described how German General Bayerlein lost his chance at Bastogne because he did not get information on the progress which his own troops had made at the village of Wardin. There is another object lesson in this same incident. The Americans who first took the village won it in a brisk fire fight which cost the enemy several tanks. Too, they were operating virtually as a detached force. There was no immediate support on their left and they had failed to make proper juncture with the team of armor which was supposed to be sustaining them on their right. Thus the danger was as if they were operating alone in enemy country and one would scarcely expect that there could be any let-down in their vigilance until they had again been joined by friendly forces.

Yet within an hour of this American capture of Wardin the enemy came back into the village with armor and the American company was driven back and scattered, with a loss of half of its strength in killed, wounded, and captured. Later it was learned that when the German counterattack struck, the Americans were in Belgian houses, searching the pantries. And these were first-class troops which had fought many successful engagements.

Unusual? Not at all! This is the normal reaction of fighting men as quickly as the danger appears to vanish, unless they are under very strong control. They tighten up when the immediate pressure rises; they relax as the immediate pressure lifts. Of themselves, they will not remain vigilant, even though they are battle-wise. The degree of vigilance depends altogether on the measures taken by their leaders. Unfortunately, the majority of junior leaders have this same tendency. Our battle records supply many incidents illuminating this fact. As an example, during the Kwajalein campaign one company of the 17th Infantry Regiment was fully engaged in an attack on the Jap naval air base on Burton Island. On its right flank, two squads of the First Platoon had moved up abreast of an enemy air shelter. The men knew that the Japs whom they had been fighting had taken refuge in the shelter. But they took no steps to finish them off with explosive charges. Instead, they stretched out on the ground around the shelter, talking and laughing and waiting for the company's left to get in motion. Half an hour passed. The private who was supposed to be covering the shelter entrance leaned against the building, eating a candy bar. From the other end of the shelter, a Jap machine gun fired wild periodic bursts toward the company on the far right flank. But the men figured the gunner was doing no real damage and so they let him continue. This was the situation when the enemy within the shelter made their last sortie. A rifleman, lying in the grass, yelled: "My God! They're coming out!"

Though the old military maxim that "the weakest point always follows success" applies with especial emphasis to the operations of minor tactical forces, it might more sensibly be rewritten that the weakest point is when the leader relaxes. This being the natural reaction of troops, there is no safeguard against it other than double vigilance on the part of those who command.

Despite the near presence of the enemy, troops will always let down at every opportunity and it is the task of leadership to keep them picked up. They will always bunch unless they are insistently told by voice to stop bunching. They will always run if they see others running and do not understand why. In these natural tendencies lie the chief dangers to battlefield control and the chief causes of battlefield panic.

On seven occasions it has been a part of my duty to investigate the sources of panic along the battle line, twice in the Pacific and five times in Europe. They were not large-scale panics but local affairs, choked off by heroic measures near the outset. But since panic gathers volume like a snowball, I think we can take it that every large panic starts with some very minor event and that for the general purposes of control it is more important to have exact knowledge of the small cause than the large effect.

By the time these seven investigations were concluded, we were able to say in each case how the panic got started and to name the persons engaged in the precipitating incident. The facts were still fresh in all minds when the hearings were held. The conditions were such that the witnesses knew they could speak the truth with impunity.

Of the seven incidents not one can be considered a "spontaneous movement" by a body of men. That is to say that they arose from conscious acts. In the second stage, in each case, there was blind, instinctive flight by a body of men, but that was not the true beginning of things. They ran as a body because something had happened which had made them suddenly and desperately fearful.

In every case this something could have been avoided. That was the common denominator, that the trouble began because somebody was thoughtless, somebody failed to tell other men what he was doing. I think it can be laid down as a rule that nothing is more likely to collapse a line of infantry in combat than the sight of a few of its number in full and unexplained flight to the rear. Precipitate motion in the wrong direction is an open invitation to disaster. That was how each of these seven incidents got its start. One or two or more men made a sudden run to the rear which others in the vicinity did not understand. But it was the lack of information rather than the sight of running men which was the crux of the danger. For in every case the testimony of all witnesses clearly developed the fact that those who started the run, and thereby spread the fear which started the panic, had a legitimate or at least a reasonable excuse for the action. It was not the sudden motion which of itself did the damage but the fact that the others present were not kept informed.

For example, a sergeant in the First Battalion, 502d Infantry, was hit through an artery during the Carentan Causeway fight on June 12, 1944. It happened in a flash. One second he was hit and the next he was running for a first-aid station without telling his own squad why he was getting out. They took out after him and then the line broke. Others who hadn't seen the sergeant make his dash saw someone else in flight. They too ran. Someone said: "The order is to withdraw." Others picked up the word and cried it along the line: "Withdraw! Withdraw!" It happened just as simply as that.

In one incident in the Pacific (reported in the book *Island Victory*) an artillery observer's party had had its radio drowned out. To continue communication the observer asked and received permission to withdraw to the company CP so that he could use the radio there. But because there was considerable mortar fire falling along the front, the party withdrew at a run instead of a walk. The men in the infantry line saw these men pass in a rush (it was night) and they got up and ran.

The third example is from the action of the La Fière Bridgehead, Normandy, June 9, 1944. An infantry company commander gave an order for a limited withdrawal. But he was with the left flank platoon and the order didn't carry to his right flank. He then began an orderly but rapid withdrawal of the left flank; the flank remained under control and stopped at the line designated—the first hedgerow to the rear. The right flank, seeing the movement but not understanding the order, promptly took to its heels. Others joined the flight as it passed groups of skirmishers and only heroic measures by a few individuals stemmed a tide that threatened the whole position.

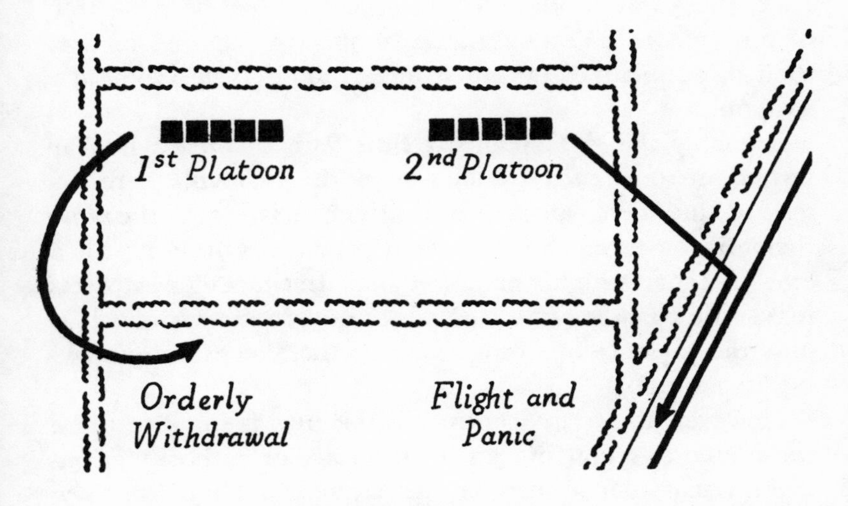

These are typical incidents. The others are quite like them. In the beginning there is precipitate and unexplained motion. Other men nearby become stampeded by the appearance of flight. Finally, unity of action is restored by the prompt decision of a few volunteers who stand squarely in the path of the flight, command the men to turn back, and do not hesitate to manhandle such of the men as come within reach or to threaten the others with weapons.

One other common denominator is worth noting. The panics were all stopped before the panicked men had run themselves to exhaustion. Had this not occurred, it would not have been possible to turn the men back to their duty. Troops which are permitted to keep going until they are physically spent are temporarily worthless as soldiers, both physically and morally. Nothing will give them tactical value until they are rested. Panic extended to the point of utter fatigue produces an atrophy of body and spirit. I have seen such panic in the faces of men during amphibious operations. The enemy confronted them and the sea was at their back; there was no place to run even had they been capable of movement. They sat there dumbly in the line of fire, their minds blanked out, their fingers too nerveless to hold a weapon.

Probably the great panic at Bull Run in the Civil War started in some such trivial way as the incidents here described, and there would be only minor variations of the same theme in most cases of battlefield panic. An army is still a crowd, though a highly organized one. In times of great stress it is subject to the same laws which govern crowds and it is only the presence of strong control which keeps it from acting like a mob.

However, the term "control" is not in this instance to be considered as synonymous with the voice of authority. Control is a man-to-man force on the battlefield. No matter how lowly his rank, any man who controls himself automatically contributes to the control of others. Fear is contagious but courage is not less so. To the man who is in terror and verging on panic, no influence can be more steadying than that he see some other man near him who is retaining self-control and doing his duty.

In the normal man it is an absolutely normal impulse to move away from danger. Yet within an army it is recognized by all that personal flight from danger, where it involves dereliction of duty, is the final act of cowardice and of dishonor. During combat the soldier may become so gripped by fear that most of his thought is directed toward escape. But if he is serving among men whom he has known for a long period or whose judgment of him counts for any reason, he still will strive to hide his terror from them.

Wherever one surveys the forces of the battlefield, it is to see that fear is general among men, but to observe further that men commonly are loath that their fear will be expressed in specific acts which their comrades will recognize as cowardice. The majority are unwilling to take extraordinary risks and do not aspire to a hero's role, but they are equally unwilling that they should be considered the least worthy among those present.

I imagine that those who are versed in the sciences would see in these statements simple proof that the ego is the most important of the motor forces driving the soldier, and that if it were not for the ego, it would be impossible to make men face the risks of battle. From that point, one could go on to say that social pressure, more than military training, is the base of battle discipline, and that when social pressure is lifted, battle discipline disintegrates. But I would prefer the simple statement that personal honor is the one thing valued more than life itself by the majority of men. The lips of the dying attest how strongly this force influences individual conduct in battle. A young company runner, hit by a shell at Carentan, collapsed into the arms of his commander, and with his life swiftly ebbing, said: "Captain, this company has always called me a — up. Tell me that I wasn't one this time." The captain replied: "No, son, you sure weren't," and the boy died with a smile on his face.

But while an army is a collection of individuals, it is also a crowd and under pressure it tends ever to revert to crowd form. The seeds of panic are always present in troops so long as they are in the midst of physical danger, the form of which changes from moment to moment. In the majority of men the retention of self-discipline under the conditions of the battlefield depends upon the maintaining of an appearance of discipline within the unit. Should the latter begin to dissolve, only a small minority of the most hardy individuals will retain self-control. The others cannot stand fast if the circumstances appear to justify flight. When other men flee, the social pressure is lifted and the average soldier will respond as if he had been given a release from duty, for he knows that his personal failure is made inconspicuous by the general dissolution. Too, it is a normal tendency in troops that they will drift rearward from the fire line unless they are being given active direction. But it is just as normal that they will reverse themselves as quickly and return to their duty if given a firm order by someone whom they know. The attitude of troops in this circumstance is usually not unlike that of a small boy caught in the act of playing hooky.

It is therefore to be noted as a principle that, all other things being equal, the tactical unity of men working together in combat will be in the ratio of their knowledge and sympathetic understanding of each other. Lacking these things, though they be well-trained soldiers, they are not likely to adhere unless danger has so surrounded them that they must do so in order to survive, and even then, quick surrender is the more probable result. But having noted the principle, it should be noted further that it is honored by the personnel system of our own Army more in the breach than in the observance. We have never had any continuity of policy which is based upon the simple idea that *esprit de corps* depends upon comradeship and our changing policies too frequently reflect an obliviousness to the factors which compose the moral strength of fighting forces.

In the defensive phase of the Ardennes operation we were given perhaps the best opportunity ever had in military operations for detailed study of the characteristics of battle stragglers. We uncovered certain traits which were common to the man power of all broken units and which are directly related to the question of how and why men can be conditioned to face fire bravely. These facts were conspicuous:

> Individual stragglers had almost no combat value when inducted into a strange organization. The majority of them were unwilling to join any such solid unit which was still facing the enemy. The minority, after being given food and a little rest, took their place in line. But the moment the new unit came under enemy pressure, these individuals quit their ground and ran rearward, or sought cover somewhere behind the combat line.
>
> On the other hand, that was not true of gun crews, squad groups, or platoons which had been routed from their original ground and separated from their parent unit, but had managed in some way to hold together during the fall-back. Upon being inducted into a strange company, they tended to fight as vigorously as any element in the command which they had newly joined, and would frequently set an example of initiative and courageous action beyond what had been asked of them.

There was scarcely a commander who fought on that ground and who had experience with both of these categories but commented to me on the absolute contrast between them. It was proved by repeated trials that the individual straggler was of such little value that it was hardly worth while attempting to do anything with him. But three or four men who hailed from the same small unit, and knew one another, would stand and fight, if welcomed into a new command. I don't believe there is any mystery about this difference, even though it is a fair assumption that all of the men had about the same moral level in the beginning, since they came from the same battalions and regiments. Within the group increments the men were still fighting alongside old friends, and though they were now joined to a new parent body, they were under the same compulsion to keep face and share in the common defense. The individual stragglers were simply responding to the first law of nature which began to apply irresistibly the moment they were separated from the company of men whom they knew and who knew them.

Of course there were other consequences of defeat which had contributed to the demoralization of these troops. It is therefore well to note that one finds the same lesson salient in the tactical studies of offensive operations by American paratroops during the last war. This has an especial significance, since paratroops must always reckon on the probability of a bad scramble during the drop and their training seeks to emphasize methods by which men can be hardened against this contingency. It is far more vital in airborne operations to have men who can pick up and go ahead confidently in any company than to have these same characteristics in regular infantry.

But as we learned by first-hand study of their operations, paratroops are rarely able to attain to this standard, and as with other troops, their battle morale, willingness, and efficiency are in the ratio of their knowledge of the men on whom they are depending for close support. Out of about seventy tactical episodes which I examined in the airborne phase of Operation Neptune (Normandy) and which were of this nature, there was only a minor fraction in which mixed forces had achieved a substantial success. It is true that almost any kind of an amalgam managed to do valiant work so long as it included fractions from various units which were working directly under their own junior leaders. But if this last characteristic was not present and if an officer or non-com, having collected a group of men whom he had never seen before, tried to lead them into battle, the results were almost uniformly unsuccessful. The men invariably stalled; the fact that they did not know the others present was to them a sufficient excuse why no action should be attempted. If the leader got them to follow until they came in contact with the enemy, at that point they were more likely to fall away from him than to support him. They would assemble readily enough under a stranger and they would usually march under him, but they would not fight for him. There were very few exceptions to this rule.

I am inclined to believe that the basic difficulty here is not so much that the leader cannot command the respect and confidence of men whom he does not know, for the brave example set by many of the very leaders who were let down by these scratch forces should have been enough to inspire a wooden Indian. It derives from the same mental block noted in the stragglers of the Ardennes—the inherent unwillingness of the soldier to risk danger on behalf of men with whom he has no social identity. When a soldier is unknown to the men who are around him he has relatively little reason to fear losing the one thing that he is likely to value more highly than life—his reputation as a man among other men.

However much we may honor the "Unknown Soldier" as the symbol of sacrifice in war, let us not mistake the fact that it is the "Known Soldier" who wins battles. Sentiment aside, it is the man whose identity is well known to his fellows who has the main chance as a battle effective.

An officer must teach his men to honor the country and the uniform and to respect the symbols of these things. But one way in which to do them honor is to begin to understand the limits in which they operate as forces influencing the conduct and shaping the fortunes of combat troops. The rule for the soldier should be that given the Australian mounted infantryman when he asked the Sphinx for the wisdom of the ages: "Don't expect too much!"

It should not be expected that pride in a uniform or belief in a national cause are of themselves sufficient to make a soldier steadfast in danger and to persuade him to give a good personal account of himself in battle even though he has lost his identity with other men and there are none to make note of who he is or whether he serves well or badly.

There are of course individual exceptions. There are men who do not need to draw moral strength from other men, who are at their best when they go it alone, who love danger for its own sake, and who become restive when life becomes tranquil. But such is not the nature of the majority of men. It is a mistaken doctrine which would have us believe that by repeating platitudes and by teaching men to snarl when going at a bayonet course, we can train them into something other than what they are by nature. Ardant du Picq wrote truly when he said: "The heart of man does not change." The chance for tactical progress is therefore dependent on (1) arriving at a clearer understanding of our human material, and (2) initiating methods which are calculated to further the growth of unity and an intellectual appreciation in all ranks of the problems which attend this principle.

Lofty ideas and ideals we must have, if only to assure that man will go forward. But it is unworthy of the profession of arms to base any policy upon exaggerated notions of man's capacity to endure and to sacrifice on behalf of ideals alone. In battle, you may draw a small circle around a soldier, including within it only those persons and objects which he sees or which he believes will influence his immediate fortunes. These primarily will determine whether he rallies or fails, advances or falls back.

Unconsciously, perhaps, we have recognized this truth. The noblest phrases in our whole military tradition pay tribute to the overpowering impact of local situation upon the spirit and will of the fighting man.

"Don't give up the ship!"

"Don't fire till you see the whites of their eyes!"

"I think it would be better to order up some artillery and defend the present location." . . . The words of U. S. Grant during the Battle of the Wilderness.

"To hell with our artillery mission. We've got to be infantrymen now." . . . The words of Lieutenant Colonel Thornton L. Mullins at Omaha Beachhead.

"If we've got to die, I don't know a better place than this." . . . The words of Capt. John J. Dolan at the Merderet Crossing.

"This is our last withdrawal. Live or die, this is it." . . . The words of Lieutenant Colonel Ray C. Allen at Bastogne.

To some extent we have recognized these same moral values in our training doctrine. We utter them in our striving for a system of discipline which recognizes in theory—though the theory is too often ignored in practice—that the relationships within our Army should be based upon intimate understanding between officers and men rather than upon familiarity between them, on self-respect rather than on fear, and above all, on a close uniting comradeship.

But even in the best circumstances the objects of the philosophy on which any discipline is based will be thwarted unless all personnel policies having to do with the re-enforcement system in wartime, officer assignments, and the preservation of the integrity of tactical units are concentric with the central purpose.

In the past this has not been done. While giving lip service to the humanitarian values and while making occasional spectacular and extravagant gestures of sentimentality, those whose task it was to shape personnel policy have tended to deal with man power as if it were motor lubricants or sacks of potatoes. They have destroyed the name and tradition of old and honored regiments with the stroke of a pen, for convenience's sake. They have uprooted names and numbers which had identity with a certain soil and moved them willy-nilly to another soil. They have moved men around as if they were pegs and nothing counted but a specialist classification number. They have become fillers-of-holes rather than architects of the human spirit.

Therein lies a great weakness, and we have suffered from it through every war. For it must ever remain so—that acquisition of a truer knowledge of the nature of man in war will suffice very little if put to work only by the local commander on limited ground. That again leaves too much to chance and puts too high a premium on the virtues and talents of the average leader.

What is needed, primarily, if we are to go forward are policies stemming from the top, which are based not upon slide-rule calculations but on knowledge of the human heart.

11

THE AGGRESSIVE WILL

"Finally, the great distinction of this game is that it truly, when well played, determines who is the best man—who is the highest-bred, the most self-denying, the most fearless, the coolest of nerve, the swiftest of eye and hand. You cannot test these possibilities wholly, unless there is a clear possibility of the struggle's ending in death."

—JOHN RUSKIN in *The Crown of Wild Olives.*

UP TO now we have considered various things such as person-to-person communication which will build the fighting spirit in troops, once they are committed to combat, and we have considered other things which are destructive of that spirit.

I believe this is the proper approach to the subject and that it should be well recognized that everything which touches the circumference of tactics bears sooner or later on the heart of the fighting man—his will to win, his courage to act and to endure. For my own part, I confess that I do not believe there is any such thing as morale, if by that word is intended some special will or purpose which can be separated from the body of the thought of an army or of one of its individuals. Either morale is a word of convenience, employed because it is much easier to say than "the body of thought of a person or persons, as to whether it disposes the thinker to high endeavor or toward failure," or else it has no meaning. I prefer to think that it is the former because we then have something to go to work on.

Morale is the thinking of an army. It is the whole complex body of an army's thought: The way it feels about the soil and about the people from which it springs. The way that it feels about their cause and their politics as compared with other causes and other politics. The way that it feels about its friends and allies, as well as its enemies. About its commanders and goldbricks. About food and shelter. Duty and leisure. Payday and sex. Militarism and civilianism. Freedom and slavery. Work and want. Weapons and comradeship. Bunk fatigue and drill. Discipline and disorder. Life and death. God and the devil.

The advantages of this new approach to a very old subject are quite obvious. For one thing, it automatically disposes of the need for any fine distinctions between "training morale" and "combat morale." Wherever the soldier may be and whatever he may be doing, his morale is still the product of his whole thought. He may arise in the morning feeling like a lion because he has rested well and his breakfast is good; two hours later he may be sagging in the middle because of ten pounds excess weight in his pack; three hours later he may have become a cipher because the socks which his mother had knitted him have worn blisters on both of his heels. Or some other man in combat may feel contempt for his commander and mortal fear of the enemy and still face the fire bravely because of the love of a woman and his own resolve not to discredit her.

Too, the definition cuts through one of the oldest myths in the military book—that morale comes from discipline. If that were true it would indeed be possible to extract the greater from the lesser. The process is precisely the reverse: whether on the field of battle or in "pirouetting up and down a barrack yard" as Carnot's phrase has it, true discipline is the product of morale. By "true discipline," however, I do not mean a mere muscular response to orders. The latter comes from the training of body and wits; the former derives from a decision of the intellect which is based largely upon self-interest. I think that one of the general mistakes made by the military body is that because soldiering is a patriotic calling, it is regarded as somehow base to put self-interest foremost in appealing to the judgment and imagination of the soldier. Yet it is undeniable that a willingness to accept the system is the first step in the soldier's personal advancement. As his success enhances his appreciation of the military life, he grows in knowledge and in a willing ability to apply it. These are the constituent elements of the true discipline upon which an army is finally dependent. The soldier needs a sound and vigorous body if he is to contend in modern war; but this itself should be the object rather than the perfecting of him in drills not even remotely related to his use of weapons, out of a mistaken and obsolescent notion that they somehow improve his discipline. There is no time for such methods. Moreover, they are out of date, since they stem from the doctrine that the soldier can be trained to respond as if by habit. Beyond the basic physical requirement, the essential is that he be given freedom to think with a clear mind, which freedom can be his only when he becomes convinced that the Army—and particularly the Army as represented by his immediate superiors—is doing everything possible for his welfare.

Soldiers can endure hardship. Most of their training is directed toward conditioning them for unusual privation and exertion. But no power on earth can reconcile them to what common sense says is unnecessary hardship which might have been avoided by greater intelligence in their superiors. The more intelligent the soldier, the more likely it is that he will see that as a sign of indiscipline up above and will answer it in the same way.

Nothing more radical is suggested here than that the leader who would make certain of the fundamental soundness of his operation cannot do better than concentrate his attention on his men. There is no other worthwhile road. They dupe only themselves who believe that there is a brand of military efficiency which consists in moving smartly, expediting papers, and achieving perfection in formations, while at the same time slighting or ignoring the human nature of those whom they command. The art of leading, in operations large or small, is the art of dealing with humanity, of working diligently on behalf of men, of being sympathetic with them, but equally, of insisting that they make a square facing toward their own problems. These are the real bases of a commander's major calculations. Yet how often do we hear an executive praised as an "efficient administrator" simply because he can keep a desk cleared, even though he is despised by everyone in the lower echelons and cannot command a fraction of their loyalty!

Napoleon was thinking of these fundamental values when he said that those who would learn the art of war should study the Great Captains. He was directing our gaze toward the manner in which Alexander, Caesar, and Hannibal had sought the keys to military success in an understanding of human nature and in the molding of its power to their tactical and strategical purposes. The leaders in our own Civil War liked to boast about their "thinking bayonets." If the term had in that day a tone of overstatement, at least we have now come abreast of it.

"No man wants to die; what induces him to risk his life bravely?" Field Marshal Sir Archibald Wavell once asked the question. The only answer which occurs to me as supportable in all that I have seen of man on the battlefield is that he will be persuaded largely by the same things which induce him to face life bravely—friendship, loyalty to responsibility, and knowledge that he is a repository of the faith and confidence of others. In his leisure moments Wavell reflected that "belief in a cause may count for much." But Major General Terry de la Mesa Allen, speaking from the battleground of Sicily, declared that men do not fight for a cause but because they do not want to let their comrades down. Where lies truth?

Men who have been in battle know from first-hand experience that when the chips are down, a man fights to help the man next him, just as a company fights to keep pace with its flanks. Things have to be that simple. An ideal does not become tangible at the moment of firing a volley or charging a hill. When the hard and momentary choice is life or death, the words once heard at an orientation lecture are clean forgot, but the presence of a well-loved comrade is unforgettable. In battle the most valued thing at hand is that which becomes most stoutly defended. All values are interpreted in terms of the battlefield itself. Yet above and beyond any symbol—whether it be the individual life or a pillbox commanding a wadi in Sahara—are all of the ideas and ideals which press upon men, causing them to accept a discipline and to hold to the line even though death may be at hand.

If any man doubts that these values have a place in hardening the resolve of an army, let him answer the question: What happens when an army loses faith in its cause? It is in fact defeated and wholly submissive to the enemy. Its will is defeated. If it can expect to receive quarter, the last reason for resistance has disappeared. Those who are interested in further pursuit of this line of inquiry might study the collapse of the Czarist Army and the insurrection of the German Navy in World War I. It will be found that the decrying of their own cause was the chief manifestation of their insubordination.

Those who respect history will deem it beyond argument that belief in a cause is the foundation of the aggressive will in battle. But that does not mean that it should become the main object in an army to turn every soldier into a doctrinaire with the expectation that it will either increase his military efficiency or his desire to do the enemy to death. If a soldier has an initial belief in the justice of the politics of his country, this belief will be nourished in the measure that the country keeps faith with him during his service. While there are many extraneous factors which enter into his emotional and intellectual judgment of these matters, none counts as heavily as the manner in which he is treated by those who are in authority over him. Though belief in the nation is the foundation of his personal discipline, the superstructure is raised by human hands which toil within his sight. There is nothing more soulless than a religion without good works unless it be a patriotism which does not concern itself with the welfare and dignity of the individual. Only the officer who dedicates his thought and energy to his men can convert into coherent military force their inarticulate thoughts about their country; nor is any other in a position to stimulate their desire to be of service to it.

Having already suggested that the thinking bayonets can never be subordinated by the routine methods of a discipline which is based largely on mechanical procedures, I will add that no leader will ever fail his troops (nor will they fail him) who leads them in respect for the disciplined life.

Between these two things—discipline in itself and a personal faith in the military value of discipline—lies all the difference between military maturity and mediocrity. A salute from an unwilling soldier is as meaningless as the moving of a leaf on a tree; it is a sign only that the subject has been caught by a gust of wind. But a salute from the man who takes pride in the gesture because he feels privileged to wear the uniform, having found the service good, is an act of the highest military virtue. It is not mere coincidence that in those line companies which achieved phenomenal success in combat during the late war, one found always the closest of working relationships between officers and men, and one found also that the salute was given with a proprietary air, as if all ranks were glad to own it. One further fact which needs to be stressed about the character of those officers whose capacity could be measured in the efficiency of their companies —while they were scrupulous in their care of their men, they were not "wet nurses." They treated their subordinates as men; they did not regard them as adolescents and they did not employ the classroom manner in dealing with them individually or in the mass. That was an important part of their hold upon men. The latter respect manliness, not maidenliness. They prize a commander the more if he looks and acts the part of a soldier, but the characteristic of a fine appearance will but betray him the sooner if he has no real kinship with men.

The characteristics which are required in the minor commander if he is to prove capable of preparing men for and leading them through the shock of combat with high credit may therefore be briefly described:

(1) Diligence in the care of men.
(2) Administration of all organizational affairs such as punishments and promotions according to a standard of resolute justice.
(3) Military bearing.
(4) A basic understanding of the simple fact that soldiers wish to think of themselves as soldiers and that all military information is nourishing to their spirits and their lives.
(5) Courage, creative intelligence, and physical fitness.

(6) Innate respect for the dignity of the position and the work of other men.

Weighing these values, we see outlined a frame of relationships between ranks wherein all things work together toward the strengthening of the aggressive will of men in combat. For the military purposes of the United States it is necessary to reject the philosophy of the German General Hans von Seeckt that: "A true military discipline stems not from knowledge but from habit." For that must be substituted the more enlightened view that the knowledge of facts constitutes experience and the product of experience is habit, which plays a more important part in war than anywhere else in life. It would appear to me to be incontestable that if the Seeckt view of discipline could of itself vouchsafe military success, then *esprit* and patriotic incentive are superfluous and had best be counted out of the motor forces which contribute to combat effectiveness. For certainly if an impulse is not based on knowledge and understanding, it can only be grounded in the fear of the consequences of disobedience. We have in our service a few antediluvian holdovers who still cling to the old ideas of discipline and who, in the words of the Archduke Albert, "in time of peace excel in detail, are inexorable in matters of equipment and drill and perpetually interfere with the work of their subordinates." But if I am incorrect in saying that these men are out of step with their times, then for the sake of consistency we should eliminate all appeal to the soldier's reason and go by the rule that the less intelligent the soldier the more ably will he serve.

We carry on from this point to an understanding of the essentials of combat moral incentive, which, together with the degree of physical hardening and technical knowledge sufficient to insure that muscle and brain will respond to the will, constitutes the true discipline of the combat forces of the Army.

Chief among these essentials is some knowledge of the national cause and a maximum knowledge of the forfeits in the battle which is being fought; in war or elsewhere the risks which the majority of men will accept are in proportion to what they know of the importance of the undertaking.

Next comes faith in the power of the company and the higher tactical commands; this last should be supported by as much knowledge of the strength of other elements as can be provided.

And last, these things must combine with confidence in leadership and with an acceptance of the basic philosophy governing human relationships within an army.

Granted that these conditions have been met, our Army should never be put under the necessity of humoring and yielding irretrievable ground to the inevitable minority of malcontents or of permitting governing principles to be influenced by voices from the lunatic fringe, even those which have been elected to Congress.

The Army's interior economy and system of relationships are different from those within the civil body from which the Army springs. The Army must continue to maintain the broad lines of distinction or else in time it will deteriorate into an armed mob wherein the voices of Babel are substituted for the one clear voice of authority. To think otherwise is to agree finally that command decisions should be submitted to arbitration boards and grievance committees should sit on the question of where and how to maneuver. As a civilian observer I saw these very things attempted in the Anarchist "Iron Brigade" during the Spanish Civil War, which tried to operate with fidelity to the principle of the absolute equality of all ranks. In the end all discipline was destroyed and all operating capacity was lost. These same things can happen to any other service which is compelled to base its discipline wholly upon a political philosophy which does not take into account the singleness of purpose of a military body.

It is my belief that there is no more hurtful doctrine put before our people today than that the Army should duplicate the arrangements which obtain within the civil society, slavishly imitating the latter's comforts, social customs, and ideas of the regulating of justice, and insisting on no higher standards of personal responsibility for its people.

For where, when one examines the civil society, can there be found adequate proof that the majority of men and women are continuing to give willingly that measure of voluntary co-operation which is essential to the attainment of great ends within a democracy and, indeed, to the preservation of the democratic way of life? One of our chief characteristics as a people is a besetting softness, coupled with a marked unwillingness to yield any personal ground for the sake of the general good. Nor can it be maintained in honesty of spirit that our standards of enforcing justice, and of requiring respect for it, have given us anything other than an unenviable reputation as a people who do not have a sufficient regard for their own law. It is in these very particulars that today so many of our people are filled with grave misgivings about the national future. They are alarmed about the weakening of the discipline of our general society. But at the same time they fail to sense a real danger in the attack upon the discipline of our national forces.

Yet this is no new issue with us. We invariably set the clock back. Every war in the national history has caught us unprepared, materially and in the spirit. Every postwar period has found us returning almost immediately to that mood of bitter and unjust criticism of the fighting services which poisons the mind of youth against service to his country and kills the volunteer spirit. "Our Army" promptly becomes "The Army" and "Our Navy" becomes "The Navy." The familiar and always popular warnings against the dangers of a growth of militarism among our people are raised anew by leaders who never having served in their country's uniform, are ever loath to recognize that those who do so may have a devotion to the nation's welfare and a love of its free institutions quite equal to their own. These fears are best examined in the light of our own history. They are best answered in the question of whether the growth of militarism has ever been, or is now, a greater danger to our people than the growth of a materialism which chokes out all patriotic incentive and withers belief in the ideal of service to the nation.

It is assumed quite falsely by many of those who in ignorance inveigh against it that service discipline has some special and sinister aspect of its own instead of being, like the discipline of any other body, those standards of conduct which will assure a maximum efficiency according to the mission. After every war the same pressures for a relaxed discipline have been exerted. The same arguments have been advanced, the same cries for reform have been made, on the basis of the criticims made by a minority of individuals. Seldom, if ever, has there been due reflection on the possibility that by insisting that all military procedures should be measured by the yardsticks of civil custom and practice the democracy unwittingly sacrifices the one chance that a determining number of its people may learn the meaning of self-discipline and the value of submerging personal choice in the interests of a larger unity. Yet that is the line of main chance if the republic is to develop greater powers of endurance through the influence of its military establishment. Men must feel themselves a part of something greater than themselves if a nation is to achieve a high destiny, if it is to create a great tradition, if it is to endure the test of centuries. Without such a feeling, that ethical sense which is the indispensable basis of obligation is lacking, and when that sense is lacking, a state deteriorates into brigandage, as St. Augustine observed, and kingdoms become tyrannies, as Aristotle pointed out twenty-five centuries ago. The Testaments verify this fact of social organization. Inquiring minds have explored this principle, great events have underlined it, in every century. To the extent that it is permitted to bear upon the individual within the fighting services it becomes a safeguard of alert and vigorous citizenship.

In my opinion—and I do not say this lightly—the fault in our disciplinary level during World War II was not primarily that the discipline of the ranks needed to be more relaxed but that the discipline of a considerable percentage of our officers needed to be tightened. For this simple reason: In so far as his ability to mold the character of troops is concerned, the qualifying test of an officer is the judgment placed upon his soldierly abilities by those who serve under him. If they do not deem him fit to command, he cannot train them to obey. Thus when slackness is tolerated in officership, it is a direct invitation to disobedience, and as disobedience multiplies, all discipline disappears.

In a thoroughly efficient army there is the essential difference between the two forms of discipline that one is preeminently concerned with command and the other with obedience. No officer can command unless he is certain of himself and confident that his orders are likely to lead to success. On the other hand, command does not create its own magic. Men who are filled with the spirit of disobedience will break the heart and ruin the character of the finest officer who ever lived. A nation which encourages its youth in this spirit cannot be saved by an officer corps, however able. And it is not likely to be able, for the same spirit will infect the corps.

Having heard many able commanders speak in frank and sincere criticism of the Army's wartime orientation program, saying that much of it was badly applied and that valuable time was wasted (a conclusion with which I am in complete agreement), I feel that it is all the more necessary to outline wherein these teachings, when supplied by men who have already won the confidence of the soldier, serve a direct tactical purpose.

The constituent elements of moral incentive, as previously outlined in this chapter, are indivisible. Each draws strength from the other. The soldier who is confident of his unit has greater faith in his country. The patriot will rally to the signs of strength in his organization and will seek ways to serve it. But while these influences are in all respects complementary and mutually supporting, men must have a well-rounded faith in the rightness and the power of all authority which is over them if a command is to survive the ordeal of heavy and successive defeats.

When confidence in the character of the unit falters, the unit itself is on the way to defeat and dissolution, without proportionate gain to the over-all undertakings of the army. Just as certainly, faith in the rightness of a cause cannot sustain the fighting will of troops unless they believe that the larger military undertakings will succeed. One may read this meaning from the great episodes of the war just past. Dunkirk, Bataan, Stalingrad, El Alamein, and Bastogne are monumental testaments to the fact that the belief in ultimate victory does more than all else to rally tactical troops and to persuade them to sell their lives dearly. In contrast, there is the example of the quick collapse of France when Frenchmen heard their leaders say that the democatic cause was lost.

The way in which the loss of moral incentive is reflected in the tactical behavior of troops was thoughtfully expressed many years ago by the British teacher, Colonel G. F. R. Henderson:

> When troops once realize their inferiority, they can no longer be depended on. If attacking, they refuse to advance. If defending, they abandon all hope of resistance. It is not the losses they have suffered but those they expect to suffer that affect them. Consequently, unless discipline and national spirit are of superior quality, and unless the soldier is animated by something higher than the habit of mechanical obedience, panic, shirking and wholesale surrender will be the ordinary features of a campaign.

When men become fearful in combat, the moral incentive can restore them and stimulate them to action. But when they become hopeless, it is because all moral incentive is gone. Soldiers who have ceased to hope are no longer receptive beings. They have become oblivious to all things, large and small.

Yet it would be most idle to conclude that in the training of the soldier there is any line of demarcation between the forces of the spirit and the strength which comes of bodily well-being. They, too, are mutually supporting. It profits an army nothing to build the body of the soldier to a gladiatorial physique if he continues to think with the brain of a malingerer. On the other hand, participation in sport may help turn a mild bookkeeper into a warrior if it has conditioned his mind so that he relishes the contest. The act of teaching one man to participate with other men in any training endeavor is frequently the first step in the development of new traits of receptiveness and outward giving in his character. It is from the acquiring of the habit of working with the group and of feeling responsible to the group that his thoughts are apt to turn ultimately to the welfare of the group when tactical disintegration occurs in battle; the more deeply this is impressed into his consciousness, the quicker will he revert under pressure to thinking and acting on behalf of the group. The reason why most infantry regiments find their headquarters people lacking in tactical integrity during a battle emergency is that they rarely bother to continue their group activity after the completion of basic training. Airborne infantry regiments do not make this error and their headquarters personnel are as efficient tactically as any other.

Among the ever-pressing problems of the commander is the seeking of means to break down the natural timidity of the great majority of his men. This he can never do unless he is sufficient master of himself that he can come out of his own shell and give his men a chance to understand him as a human being rather than as an autocrat giving orders. Nothing more unfortunate can happen to the commander than to come to be regarded by his subordinates as unapproachable, for such a reputation isolates him from the main problems of command as well as its chief rewards. But when his reticence is due to his own timidity rather than to mistaken ideas about the attitude which profits most in his post of responsibility, the only cure for him is to dive head-first into the cold, clear water, thereby growing in self-confidence even as he progresses in knowledge of the character of his men and of human nature in general.

As a further safeguard against making himself inaccessible, the commander needs to make an occasional check on the procedures which have been established by his immediate subordinates. At all levels of command it is not infrequently the favorite self-appointed task of these people to think up new rules which will keep all hands "from bothering the old man." Many a likely and likeable commander leads an unwontedly lonely life because of the peculiar solicitude of his staff in this matter and his own failure to discover what is happening to him. In this way the best of intentions may be thwarted.

It is never a waste of time for the commander to talk to his people about their problems; more times than not, the problem will seem small to him, but so long as it looms large to the man, it cannot be dismissed with a wave of the hand. He will grow in the esteem of his men as he treats their affairs with respect.

There ought to be a way to underscore this principle, for I have found it a common fault in our younger officers that they do not know how to talk to their own men. The approach too often has an air of condescension or of forced restraint. This tendency toward a kind of intellectual grade separation between officer and man is a block to progress in all operations, though it is less often due to priggishness than to lack of instruction and careful coaching. Whatever the cause, aloofness on the part of the officer can only produce a further withdrawal on the part of the man. Finally, the cost comes high. In battle, and out of it, the failure to act and to communicate is more often due to timidity in the soldier than to fear of physical danger.

To the young officer who is conscious of his own reserve and is anxious to do something about it, I can suggest nothing better than to make a habit of full physical participation. That is, instead of watching the squad or platoon work out a problem and either directing or criticizing its action, let him pick up a weapon, relieve one man in the group, then let himself be one of the commanded until the conclusion of the operation. Is this course of conduct beneath the dignity of an American officer? Certainly not! It was Major General Percy W. Clarkson's method of making personal reconnaissance a moving force in the operations of the 33rd Infantry Division during World War II. He did not preach to his officers about the value of personal reconnaissance. But it was his habit when his regiments moved out on night problems to appear quietly at the scene, fill a blank file in one of the rifle squads, and remain with the squad until the problem was concluded. The effect was twofold: the men took greater interest in their work and the officers formed the habit of getting down to the men.

While the subject of physical conditioning for war is far outside my special sphere, it is appropriate to comment here at least that when troops lack the co-ordinated response which comes only through long and rigorous training, their combat losses will be excessive, they will lack cohesion in their action against the enemy, and they will uselessly expend much of their initial velocity. In the United States service we are tending to forget, because of the effect of motorization, that the higher value of the discipline of the road march in other days wasn't that it hardened muscles but that, short of combat, it was the best method of separating the men from the boys. This is true today, despite all the new conditions imposed by high velocity warfare. A hard road march is the most satisfactory training test of the moral strength of the individual man.

The great advantage of the gain in moral force through all forms of physical training is that it is an unconscious gain. Will power, determination, mental poise, and muscle control all march hand-in-hand with the general health and well-being of the man. Fatigue will beat men down as quickly as any other condition, for fatigue brings fear with it. There is no quicker way to lose a battle than to lose it on the road for lack of adequate preliminary hardening in troops. Such a condition cannot be redeemed by the resolve of a commander who insists on driving troops an extra mile beyond their general level of physical endurance. Extremes of this sort make men rebellious and hateful of the command, and thus strike at tactical efficiency from two directions at once. For when men resent a commander, they will not fight as willingly for him, and when their bodies are spent, their nerves are gone. In this state the soldier's every act is mechanical. He is reduced to that automatism of mind which destroys his physical response. His courage is killed. His intellect falls asleep.

Truly then, it is killing men with kindness not to insist upon physical standards during training which will give them a maximum fitness for the extraordinary stresses of campaigning in war. As the body is hardened, so must the mind be steadily informed, so that the soldier will take a reasoning view not only of the privations of the field but of that which is being attempted. Once we depart from the ideal of automatic response as the condition most likely to produce unity of action in battle, the only substitute for it lies in the possibility that more and more men in the ranks can be trained to see and think through the haze of battle in unison with their commanders.

That being the case, it is my belief that this whole subject of the will of the commander cries out for a modern resurvey and better understanding. That such a genius as Clausewitz avoids discussion of the area within which the will operates is not strange, the nature of war in his time considered. But it is remarkable that moderns like Ludendorff, Fuller, and Wavell have been similarly vague about it. Invariably, the importance of the will of the commander is discussed as if men were like oxen, seeing but not thinking, waiting for food or for the goad, incapable of holding others to account.

To further any such idea will not be conducive to the best results, for it is important that the young commander see all things in balance. Simply to exalt the will as a force without limit tends only to create delusions about what may be accomplished by the exercise of it. These delusions in turn destroy the confidence of troops, then as confidence flags, they become unresponsive to the bidding of the commander. The quality of mind most worth seeking is not power of will but, as said by Marcus Aurelius, "Freedom of will and undeviating steadiness of purpose."

But it is a good question: Why is the will of the military commander deemed more decisive of success than the will of leadership in any other calling? Clearly it is because the inertia, frictions, and confusions of the forces of the battlefield make all positive action more difficult. And yet the principles which win intelligent man-to-man co-operation apply equally in all circumstances. The same rules work on the battlefield as in an office.

The will does not operate in a vacuum. It cannot be imposed successfully if it runs counter to reason. Things are not done in war primarily because a man wills it; they are done because they are do-able. The limits for the commander in battle are defined by the general circumstances. What he asks of his men must be consistent with the possibilities of the situation.

What can be successfully willed must first be clearly seen and understood. If amid the confusions of battle the commander sees what is required by the situation, if amid the miscarriage of arrangements and the assailing doubts of other men he measures the means of doing it, and if he then gives his order and holds his men to their duty, this is the ultimate triumph of the will on the field of battle. To reflect on this thought is to note that he exercises his will far less upon his men than upon himself.

Should he, on the other hand, attempt to will that which his men know cannot be done or feel unanimously is utterly beyond reason, or should he base his order on assumptions which they recognize as false, his will becomes temporarily without power and cannot help the situation.

There is scarcely a commander with any time in combat but has had this experience of having willing troops become suddenly unresponsive because the facts were not straight. It happened to Lieutenant Colonel H. W. O. Kinnard during the night advance of the First Battalion, 501st Parachute Infantry, against the German-held Dutch village of Schijndel in September, 1944, which I have praised elsewhere in this book as one of the most brilliant battalion actions of the war.

The column had advanced only five hundred yards beyond its outpost line when it came under machine-gun and anti-tank fire. Kinnard heard some of the fire clipping the branches of the tree above his head and judged it was all going high. But the lead company had stopped and the men had jumped for the ditches. He ran forward, shouting to the men: "Keep going! Keep going! That fire is high." But his personal advance had no effect. Not a man stirred. Within a moment he understood why he had failed. From one of the ditches a rifleman answered him: "If you think the fire is going high, Colonel, come over here. We've had eight men hit in the legs."

That is a slightly dry but pointed example of the will failing for natural causes, and perhaps its negative lesson is as instructive as is the familiar and triumphant story of Napoleon and the Toulon battery.

In war, the will may triumph only as it is the expression of massive common sense, conditioned by an accurate appreciation of the general emotional situation. Thus it is without virtue except as it is applied knowledge of the situation and of human frailties and capabilities under varying conditions. Wherein it stems not from knowledge but from ignorance, it is obstinacy, which is one of the worst of faults in any soldier. There are some things that cannot be done in war. For example, troops cannot operate in temperatures of 50° below zero, no matter how warmly they are clad, and at 25° below they cannot fight more than thirty minutes or so in a given hour and still remain tactically useful. So far as we know, these natural limits may not be overcome by any will or wit of man.

As to the training of the will, the primary rule was given by the great general, Marcus Aurelius: "Attend to the matter which is before thee, whether it is an opinion or an act or a word." For power of will comes from the exercise of it. In the military commander strength of will does not automatically flourish apace with the growth of knowledge in the material things of war. But unless it does so, experience is had to no broad purpose and knowledge does not become wisdom. Equally, if the will power of the commander is in excess of his knowledge, it cannot be exercised for the good of the commanded. Thus, true strength of will in the commander develops from his study of human nature, for it is in the measure that he acquires knowledge of how other men think that he perfects himself in the control of their thoughts and acts.

It has been the misfortune of armies that in all times past the importance of the will has been discussed always *by* generals in terms which would suggest that it applies chiefly *to* generals. Such, surely, is not the case. The good general is simply the good company commander in his post-graduate course. The idea that more godlike qualities are required of him and that he above others can achieve miracles through the working of his will is dismissed as idle superstition. It is in the small fight that the will is hardened by experience and tempered by the lessons of experience. It is in the skirmish that the young commander is most likely to learn that the thing which is willed must first be seen and that the test of will lies not in the giving of an order but in the measuring of the stituation upon which the order is based.

The major battle is only a skirmish multiplied by one hundred. The frictions, confusions, and disappointments are the same. But there is this absolute difference—that the supreme trial of the commander in war lies in his ability to overcome the weaknesses of human nature in the face of danger, and these are matters which he cannot know in full unless he has served with men where danger lies.

Missing that experience, or availing not of its lessons, the commander, in de Saxe's words, "Remains dependent on the face of fortune, which in war is most inconstant." He makes men the servant of his will rather than making his will the servant of man. In default of knowing what should be done, he does only what he knows.

12

MEN UNDER FIRE

"Every purpose is established by counsel and with good advice make war."
—THE BIBLE.

TO THOSE who have known the firing line, it would scarcely be necessary to point out that morale in combat is never a steady current of force but a rapidly oscillating wave whose variations are both immeasureable and unpredictable.

It is in this respect chiefly—the rapidity and capriciousness of its variations—that the morale problem in the zone of fire differs from that of rear area soldiering.

A band of men may go through a terrible engagement, take its losses bravely, and then become wholly demoralized in the hour when it must bury its own dead.

A regiment, fretted to utter abjection by a protracted stay in the lines, may find its fighting spirit again in a six-hour respite during which the men are deloused and given a change of underwear.

A battalion, advancing boldly, may be brought in check because its commander did it the disservice of going too far forward and getting himself killed within sight of the ranks.

A platoon may charge and capture an enemy-held hill, losing half its numbers in so doing, then run down the hill again because one of its own artillery shells landed too close and hit one man.

Such are but a few of the curious currents within the ever-moving tide. In battle there is very little order. Many times the course of events is shaped by purest accident and much that one witnesses does not seem to make sense. Yet there are such notable exceptions that it can scarcely be doubted that most of the rules for the conduct of troops under fire are sound and that much of the failure resides in man's failure properly to apply them. This was the way the eyewitness described the advance of the Second Battalion, 502nd Parachute Infantry, against the German position in Best, Holland, in September, 1944:

> The Dutch had been haying. The fields ahead were covered by the small piles of half-harvested hay. Such was the only cover in this perfectly flat terrain. From left to right the line rippled forward in perfect order and with perfect discipline. Each group of two or three men dashed on to the next hay pile as it came their turn. So confident was the movement that an observer from afar might have thought the piles were of concrete. But the machine-gun fire cut into them, sometimes setting the hay afire, sometimes wounding or killing the men behind the hay. These misfortunes stopped hardly any but the dead and the wounded. One man went down from a bullet. I heard someone yell, "Sergeant Brodie, you're next!" Another man behind the hay pile answered, "Brodie's dead, but I'm coming," and he jumped up and ran ahead. It was like a problem worked out on a parade ground. The squad leaders were leading. The platoon leaders urged them on. Those who kept going usually managed to survive. The few who tried to hold back were killed.

There are many epsiodes from the late war which equal this one in gallantry. Its unique quality lies in the fact that with an entire nation underarms and with thousands of our well-trained infantry units engaging the enemy, we would search in vain for other examples illuminating the possibility of achieving almost perfect order in infantry maneuver under modern conditions of fire. This is the only clear example in my own experience or which has been called to my attention. There were other movements in this same campaign (Market-Garden) which achieved tactically a great deal more. But there was no other advance with the same clarity and precision.

I have sought the answer to why this is so, for it would appear clear that this is the thing which we most need in combat and most rarely get. Perhaps some part of the answer lies in certain of the dominant conditions of the field which I have already described; there is a tumult in men's minds which makes them oblivious to much that they have learned in training for their own well-being in battle and it is idle to believe that there is any remedy which will ever wholly prevail against it.

At the same time, it might be well to consider the paradox that troops are truly prepared to establish order on the battlefield only when, in the course of intelligent training, they have been well forewarned of the kind of disorder they may expect there.

To do this—to make men knowledgeable of human nature as it is and as it reacts under the various and extreme stresses of the field—cannot be regarded as destructive of confidence unless it is already conceded that confidence is a false virtue.

How many times on the field of battle one sees a young commander unnecessarily dismayed and shaken because the reality is so unlike what he had envisaged! Viewing the chaos, the litter and the inaction, he thinks them the tokens of defeat because his nerve has not been steeled or his eye trained to look for the signs of order and of progress amid the confusion. It is far better that he have the truth and that it be given him gradually so that it can be seen with the whole mind and in perspective. Were that done, then out of an expectation of disarrangement he would face combat with that singleness of purpose of the skipper of a storm-tossed vessel, concentrating on his instruments instead of brooding about the buffeting which is being taken by the ship. It is this kind of confidence which is required in the good leader in minor tactics. But it was never more difficult to acqire and retain than in modern war, where ranks are no longer stimulated by shoulder-to-shoulder formations or buoyed by the knowledge that the ordeal will be of short duration.

As the events of contact and collision move men in battle, playing upon their fears and hopes, tricking their imaginations, inviting and then repelling their initiative, confronting them quite suddenly with unexpected prospects of success, dashing these prospects through some queer prank of fate, reminding them that they are mortal and at the same time stimulating their brute instincts, the same group of soldiers may act like lions and then like scared hares within the passage of a few minutes.

There is no such person as the soldier who is dauntless under all conditions of combat. There is no such unit as the company that stays good or the company that is shock-proof; there are only companies which are more resolute than others and less likely to break in the face of unexpected emergency or surprise. Marshal Maurice de Saxe wrote of the strange misadventure of a French regiment which after driving the enemy from the field, continued in pursuit. The regiment had to divide in passing a wood. As the two flanks swung round the wood on the far side, each saw the other, and mutually mistaking what they saw for a new and unexpected body of the enemy, they fled from the field as fast as they could run.

The story is novel only because it appears in a great soldier's memoir. These lightning emotional changes are common wherever men fight. On a more limited scale, the incident is repeated a dozen times in any day of battle.

In June, 1944, Lieutenant Woodrow W. Millsaps led a patrol from Hill 30 down to the causeway leading to Chef du Pont. It was a night action during operations along the Merderet River in Normandy. Though the men had volunteered for the duty, all were in a peculiarly desperate state of mind before the start. For three days the battalion had been surrounded by enemy forces; within the perimeter men were dying for want of plasma and the wounded were suffering more acutely because of the lack of water; their cries of distress had so demoralized the able-bodied men that Millsaps welcomed the chance to get away from the hill.

At the foot of the hill an enemy machine gun opened fire on the patrol but the bullets went high. The men broke and "ran like dogs." Millsaps and a sergeant beat them back with physical violence. After they were again collected, Millsaps lost almost an hour, alternately bullying and pleading with them before they would go forward.

At last they charged the enemy, closing within hand-grappling distance. The slaughter began with grenade, bayonet, and bullet. Some of the patrol were killed and some wounded. But all now acted as if oblivious to danger. The slaughter once started could not be stopped. Millsaps tried to regain control but his men paid no heed. Having slaughtered every German in sight, they ran on into the barns of the French farmhouses where they killed the hogs, cows, and sheep. The orgy ended when the last beast was dead.

Millsaps then called for volunteers to outpost the position and accompany him across the river. All but one man held back. Millsaps started on with this lone willing soldier. The man soon began to fall behind. Millsaps asked: "Something wrong with you?" He answered: "I don't think so." Then Millsaps stripped the man's jump jacket away and found six bullet holes in his upper right arm and shoulder; the soldier had not been aware of his wounds until that moment. The soldier collapsed. Millsaps continued on alone.

This is a starkly brutal incident. Its source is my personal file of battle studies wherein all such combat fragments are fitted into the tactical frame of general operation. But I would hardly describe it as a typical incident, for the study of battle yields no such thing. In battle the unusual is met usually and the abnormal becomes the normal.

The near presence of death and the prospect of meeting it at the next moment move men in many curious and contrary ways. Many men seem to change character under the guns. The life of hardship and of danger gives new strength to the truly strong and greater weakness to the truly weak. The erstwhile fatalist may become the most superstition-ridden man in a company. The dreamer may become a realist. The backward man may come forward. The arrogant may develop new bounds of compassion. The "natural-born leader" may lose his vital spark. All men who have known battle have seen such strange transformations. "Courage," Major General John W. O'Daniel has said, "is an inherent quality, but it remains an unknown quantity until all of the chips are down."

Yet if all I have said thus far conveys an idea that all soldiers should be trained to believe that combat is a life of incessant strain and hardship, it is an impression which requires full correction. Those who write romantically of war tend ever to accent the boredom, the tiresome routine, and the endless waiting which are supposed to possess at least 90 per cent of the operational period. The modern school of war correspondents seems to regard it as a duty to make the public cry over the hard lot of the soldier. The so-called realists of war fiction such as Eric Remarque view through a glass darkly every last motion of the combat soldier. But what normal man would deny that some of the fullest and fairest days of his life have been spent at the front or that the sky ever seems more blue or the air more bracing than when there is just a hint of danger in the air?

Many things which happen on the fighting line are pure melodrama and some of its episodes are the richest comedy. It is a physical law that next to where the light is highest will be found the most intense shade; man enjoys laughter most in his moment of relief from terror or from tears.

On the "Island" along the Neder Rijn front, Holland, in October, 1944, a phenomenal percentage of cows and sheep was being hit by "enemy" shellfire at night, with resultant gain in fresh meat to the American messes. One morning I visited a certain artillery and found the members of Able Battery in a particularly truculent mood. The night before they had staked out a heifer a few rods from the foxholes and had arranged for it to be hit by an "enemy" shell. The explosion duly came. There was one last moo of distress from the animal. Able Battery sprinted out into the dark to collect its steaks. "And when we got there," said a sergeant, "Baker Battery was already sitting on the animal and six of the bastards had their skinning knives out."

It is not the least of his fighting assets that the American soldier has a sense of humor which can survive the shock and strain of engagement and can make a battle stand still. I have seen that very thing happen.

On the third day of the fight for Kwajalein in Feburary, 1944, a mop-up detail from the 7th Engineers found the Jap paymaster's office and blew the safe. The haul was about 300,000 yen. The engineers sent word to 32nd Infantry, whose front line was about 200 yards forward of the spot, that it was payday, and to come and get it. The message was relayed from squad to squad right down the fire line. The infantrymen outposted their position and crawled rearward. For an hour they filed through the paymaster's office. An engineer went through the motions of getting each man to sign as he was paid in Jap cash. Then the infantry returned to the fight.

As an example of combat discipline this story is perhaps shocking to the sensibilities of any who hold that steadfastness in the offensive is the only criterion. But, fortunately, the 7th Infantry Division was led by men who could relish a jest in the grand manner and who saw in this incident that touch of human nature which makes all men akin. The man who later came to command the regiment said of it at the time: "When troops can do that, they show to each other that they have conquered fear." He was proved right that night. The Kwajalein position was already overextended when the engineers hit the jackpot. Dark fell three hours later. The infantry battalion which had lined up for pay that afternoon bivouacked with enemy forces on their front, flank, and rear. Even within their perimeter there were groups of Japanese in hiding. The position was counterattacked fiercely throughout the night but the battalion didn't yield one yard of ground.

Possibly the big difference between the basic discipline of the front and rear areas could be best summed this way: when the chips are down, the main question is not how you go about your mission but whether it succeeds. The leader who depends for success on the strength of his personality soon learns that in combat the discount rate is very high indeed. Manners and appearance continue to command respect for the individual only when he is capable of carrying his proportionate part of the burden. The criterion of command is the ability to think clearly and work hard rather than to strike attitudes or accept disproportionate risks.

The small unit commander who practices self-exposure to danger in the hope of having a good moral effect on men, instead, frays the nerves of troops and most frequently succeeds in getting himself killed under conditions which do no earthly good to the army. Troops expect to see their officers working and moving *with* them; morale is impaired when they see that their leaders are shirking danger. But they do not care to see them play the part of a mechanical rabbit, darting to the front so as to tease on the hounds. In extreme emergencies, when the stakes are high and the failure of others to act has made the need imperative, such acts are warranted. But their value lies largely in their novelty. A commander cannot rally his men by a spectacular intervention in the hour when they have lost their grip if they have grown accustomed to seeing him run unnecessary risks in the average circumstances of battle.

Again for the small unit commander, there is need to seek the true meaning of the counsel insistently given by Major General Stuart Heintzelman, Major General Frank A. Ross, and others: "Anticipation is 60 per cent of the art of command."

But like ten years in the penitentiary, it is very easy to say: "Anticipate!" and very hard to do it. This is especially true of the commander who is about to take his troops against the enemy. The layman imagines vaguely that the job is well-nigh done when, after due consideration of the ground, the commander decides by what general paths his men are to move against the enemy. Yet this is the lesser portion of his responsibility.

In line with the principles of security, mobility, and offensive action, which are always closely related to tactics, he must decide on how to constitute his reserve and how to think about it in terms of his personal relation to the battle. A reserve is always a tactical base of operations, a fulcrum to work from, a chief tool in the hands of the commander. In the offensive its chief value is not that it comprises a reservoir of strength against unexpected pressure, but that bereft of it, the commander has no general insurance against stagnation in his own action. It is often a habit with the weak commander to commit his reserve as quickly as possible and then appeal for help from the outside, using the fact as proof that he has done all possible. It is the habit of the strong commander to withhold his reserve until it is clear beyond doubt that the assault force has lost momentum at the vital point, and then employ it either as a relief or as re-enforcement to give fresh impetus to the advance.

But again on the subject of anticipation, the greater part of the commander's work revolves around support and supply planning. The following check list is not rounded out, but it states some of the points which are requisite.

(1) Estimating what support fires are needed.
(2) Double-checking to see that these fires will be provided.
(3) Checking shortages and resupplying weapons and munitions within the command.
(4) Tying in with the flanks so that they will have a clear concept of the maneuver and can estimate what support they can provide it (a duty which cannot be left to the higher level if the commander wishes to be sure on all points).
(5) Making sure that the line of supply to the rear is manned and working.
(6) Working through the next higher level to see what reserve strength will be available in an emergency.
(7) Double-checking so that all support elements will understand their points of rendezvous (a most frequent cause of maneuver going awry).

While these are all matters for the commander's anticipation, the administration of them will continue to require the greater part of his dutiful attention throughout the period of attack. Coupled with the maintenance of communications, they comprise the main problem. It is by these means that the commander moves to keep casualties low and mobility and success high. His capacity for command in battle is proved at last by his effectiveness in maintaining balance in his own fire power while tying in all outside help which may further his fortunes. Should he devote his attention too exclusively to the interior working of the company, he will fail for not having exploited his exterior resources.

If the state of a command is such that the commander's hands are not free for the administration of his exterior tasks, it is an indication that the morale and leadership of the unit are based on incorrect principles and that insufficient authority has been delegated to subordinates. There can be no more miserable cause of failure though it is a weakness that is found in many conscientious officers. The man who cannot bring himself to trust the judgment and good faith of other men cannot command very long. He will soon break under the unnecessary strain he puts on himself. Sleeplessness, nervous irritation, and loss of self-control will be his lot until he is at last found totally unfit. The ideal relationship between a commander and his subordinate is nowhere better illustrated than in a passage from the letter of instruction wherein Grant told Sherman to proceed to the destruction of Johnston's Army: "I do not propose to lay down for you a plan of campaign: but simply to lay down the work it is desirable to have done and leave you free to execute it in your own way."

The principle applies with equal force to the command of small units. A company commander can no more hope to supervise directly the acts of several hundred men in battle without reposing large faith in his lieutenants than the general of an army can expect good results to come of by-passing his staff and his corps commanders and dealing directly with his divisions. But there are generals who have failed because they did not learn this lesson well when they were captains.

One must recognize, however, that there is an additional

psychological hazard. It is quite normal for the commander who is experiencing battle for the first time, no matter what his rank or length of training, to devote most of his time to worrying about the welfare and conduct of his men instead of continuing to ply himself with the questions: "Am I at my right place? Am I doing the right things?" This is a perfectly human trait. Indeed, it would be commendable if it were not excessively costly. For it can be said without qualification that the commander who so concerns himself will not have time to organize a co-ordinated attack and supervise its proper execution. Most young commanders learn this by being bruised their first time in battle or else they never learn it and continue to waste men's lives. But the ounce of prevention is worth the pound of cure. It needs to be impressed on the young officer's mind throughout his training that this will be his normal urge, once he engages the enemy, and that he must be prepared to resist it stoutly if he is to command soundly.

All I have said here should make clear that action requires an abrupt change in attitude on the part of the commander. Prior to combat the touchstone of his success is the interior working of the company; it requires the maximum of his attention. He enlarges his ability to command by advancing his knowledge of the character and potential of his men and by encouraging his lieutenants to do likewise. When he fights, he does an about-face. He must depend on his lieutenants to direct in detail the action of the men and he looks to them for much of his information essential to the regulating of the tactics of the unit as a whole. He gives orders after considering the object and the means; he then leaves the execution of the orders to other men. His own view and action must be directed primarily toward the horizons of operation.

It is his task to make certain of juncture on his flanks and to further the flankward flow of all vital information. An imposing task! In fact, it has been the practice in many of our tactical units simply to assign the maintenance of contact to whatever individual happens to be deployed on the far flank. I have seen, advancing in line, regiments which committed this difficult mission to the flank companies, which in turn committed it to the leaders of their flank squads, with the almost invariable result that contact failed in the crisis of action. One can hardly imagine a greater carelessness in battle procedure.

Co-equal with the security of flanks, the maintenance and full use of the line of communications to the rear are of major concern to the commander. It is his responsibility that the incoming supply is equal to the needs of his deployments and that the supporting arms and fires which have been promised him keep their engagements. Or if they do not, then he must raise hell about it.

Nothing will throw an infantry attack off stride as quickly as to promise it support which is not precisely delivered both in time and in volume. The effect is equally bad if they are pledged twenty tanks and only get ten or if they are promised one hundred rounds of artillery fire at 1010 and get it at 1046. The men squat in their foxholes and count. If they see a default anywhere, they feel this gives them a moral excuse to default in their portion. They procrastinate and they argue that since the promised help has not arrived, the attack is not timed to go. It takes the edge off men and creates uncertainty in their officers, with the result that the infantry attack goes off half-heartedly.

On the other hand, an artillery fire which is promptly delivered or an armor which advances steadily and confidently is like a shot in the arm. It moves the men mentally and sometimes bodily, thereby breaking the concentration of fear. But it is wiser to promise nothing than to default on anything. The memory of a default always lingers and the men consider it as proof that the higher command is letting them down.

It is not until all steps have been taken to tie in the sources of support from flanks and rear that the commander is truly prepared for the "bump" that will stop the action of the unit somewhere along its course. The stoppage may be due wholly to the fact that the unit has encountered a superior fire and is attacking strength which is beyond its means. Once again, however, I would point out that the accurate measuring of local pressure is not possible unless there is a steady flow of information from flank to flank and unless command at the higher level is aided by a staff which keeps moving up to the fire fight to reconnoiter the situation personally. All too frequently the platitude about "leaving the commander to his own fight" is so wrongfully interpreted that it shackles the thinking of the higher commander, stifles his urge toward closer searching of the action at the lower levels, and prevents him from allocating support to the area where it is most needed. Yet the line of duty would appear to be plainly marked. I have yet to meet the small unit commander in combat who didn't welcome help from any quarter or who was so cold toward his own battle that he wasn't stimulated by the coming forward of anyone who might assist him. This search is the moral obligation of the higher level; it is the only means of preserving the integrity of the command. The soldierly pride of the very best captains will many times restrain them from making any appeal until they have exhausted all interior resources and have so used up their men that the velocity of the attack cannot be restored.

When a company is stopped by physical shock, restoration of its movement becomes a problem for the battalion. When it is stopped by psychological shock, the continuing of the advance remains the problem of the company commander.

The difference between the two situations is usually one of relative losses.

Diffusion of the company over too wide a sector, a retrograde movement by supporting weapons (such as armor), the death of a well-loved officer, light losses from friendly supporting fire, enfilade fire, the appearance of some new and unexpected weapon in the enemy sector—such are a few of the causes of psychological shock. The local treatments are as many and as varied as are the diseases.

With respect to the effect of "friendly fire" hitting among troops, however, it is to be observed that if the circumstances leave any room for doubt as to the source, the men will jump to the conclusion that they are being victimized by their own guns. This impresses them as a reasonable excuse for not proceeding. The surest cure is to remove all doubt, which is best done by the sight of their own artillery putting on a good shoot out ahead.

In instances where it is unmistakable that they have been hurt by their own fire, however, the commander is ill-advised to lie to them. They will usually learn the truth later on, and when they do, it strikes a blow at his prestige. The experienced combat soldier knows that such occasional accidents are a part of battle and he accepts them as such. But he cannot make any good adjustment to the realization that his commander is either a fool or a liar. No gain ever comes from being slick with troops, from acting deviously instead of forthrightly, from posing as having superior knowledge, or from being secretive or discounting the common sense of the majority.

The most common cause of psychological shock, however, is a partial victory. The adage that "the weakest point follows success" is a fundamental truth of minor tactics and the danger is always greatest when the success is easily won. In combat the ever-present tendency is to go off balance, to fall flat on one's face when the door opens too suddenly, to fail before the second hurdle because of a careless recovery after surmounting the first.

Success is disarming. Tension is the normal state of mind and body in combat. When the tension suddenly relaxes through the winning of a first objective, troops are apt to be pervaded by a sense of extreme well-being and there is apt to ensue laxness in all of its forms and with all of its dangers.

> At Omaha Beach only one infantry company had the good fortune to get to the sea wall almost intact, with no losses of man power and nearly all equipment in hand. But having gained this first point, the company hesitated. Someone (not the commander) shouted: "Move to the right!" as it had been noted that the company had landed several hundred yards to the left of where it was supposed to come ashore. Some of the men strayed off and were hit by flanking fire. Others followed and became mixed with strange units. Within a few minutes of its first success the company was reduced to a cipher and did not contribute one thing to the tactical gains of the day.
>
> Another company in the same regiment landed even farther to the leftward of its designated sector. It gained the sea wall only after bitter struggle. The commander knew he was in the wrong area. But he made his decision instantly and ordered his men to attack straight forward up the bluff. The company forged ahead. It continued to take losses but it remained in balance and under control. At the close of day it had made one of the deepest penetrations along the Omaha Beachhead.

Clearly there is a contrast here which yields up several morals—the importance of decision, the value of a command clearly given, the gains which come of initiative, or the simple admonition given by Frederick the Great: "To advance is to conquer." After I had presented the facts to the Division Commander, he saw fit to circulate them to all troops as an example of how bold execution may atone for the initial miscarriage of a plan while uncertainty will unhinge the most promising arrangements.

But the troops themselves said they had expected to encounter their worst trouble in crossing the beach and when they reached the sea wall without being hurt, the men started to act as if they were at a clambake. They were shocked by an early appearance of quick success.

One witnesses this same moral undulation in parachute infantry after they have survived the ordeal of a night drop in the presense of the enemy. The first feeling of aloneness yields at least a dividend of extra caution, usually combined with movement, for normal men will move through danger to seek their fellows rather than remain alone. Confidence rises as the men come together. Then the coming of daylight completes the illusion of success. The men are buoyantly cheerful but they do not want to go forward and they resent being reminded of the realities of their situation. This is the lowest hour of their collective will and it requires more stubborn work by their officers to attain unified action at this point than in any subsequent stage of battle. Perhaps a medical man would explain these phenomena in terms of the working of the adrenals, but our concern is not so much the absolute cause as the tactical effect.

Nor is it only the first hurdle which entails this special danger. A certain degree of disorganization attends every stage of an advance. Immobilization is ever the child of disorganization. Troops start an attack in a certain formation; movement and enemy fire gradually destroy its integrity; unity of action becomes dissipated at the same rate. The result is that unless the commander is thinking always beyond the immediate objective and planning the means by which he can restore impetus after the object is won, the attack will bog down even though the unit hasn't suffered critical losses in man strength.

One of the most striking illustrations of how the company commander is ridden by this problem is provided by the attack on the Kergonant Strong Point which held up the advance of the 29th Infantry Division for ten days during the siege of Brest. The strength of this small fort was gradually reduced by a series of outflanking movements. On September 8, 1944, the position was directly assualted by Company G of the 115th Infantry.

The task was truly formidable. The Company had to charge 700 yards across an open field to close on the enemy works. It made this move at one bound, with the men firing from the hip and shouting the division battle cry as they ran. A few were cut down, but the others continued the run until they flopped down next to the hedgerow which covered the enemy's outer fire line. The charge lasted five or six minutes. It was executed with such dash and vigor that most of the members of the enemy garrison fled their works and escaped via a sunken road to the rearward.

Yet having almost clinched their victory in one superb effort, the Company remained immobile outside the hedge during the next seven hours. For all that time, its forward elements were within 15 to 50 yards of the few remaining Germans. Commenting on his experience of that day, during which he exhausted every resource in getting the Company moving again, the young commander, Lieutenant Robert W. Rideout, made a general summation of the causes of inertia in the infantry attack. I regard his words as especially significant because they were spoken before the assembled Company while we were yet on the ground at Brest.

> You have a plan. You have an objective. Your men get started with the objective in mind. But in the course of getting to the objective and taking up fire positions, disorganization sets in. The men look for cover and that scatters them. Fire comes against them and that scatters their thoughts. They no longer think as a group but as

> individuals. Each man wants to stay where he is. To get them going again as a group, an officer must expose himself to the point of suicide. The men are in a mental slump; they always get that way when they have taken a great risk. The junior officers feel much the same way; they see no reason why they should take an extraordinary risk. They know the men will work well when well led. But they also know that if the leaders are killed, that is the finish of the operation. The consequence is that the farther you go, the more difficult all movement becomes, provided you are moving against fire. It is harder to get men to mop-up after a charge than to get them to charge.

It is illuminating tc follow Rideout through those seven hard hours. He wasted no time prodding his lieutenants or trying to jockey his men. Once the charge had been made, he knew that they were through unless he could restore momentum through the impact of some outside agent at the vital point. It was vain to think about artillery or the heavy mortars; the lines were much too close together.

Rideout called the Company on the right flank and described his situation; the right flank Company advanced about three hedgerows. It was then flattened by fire and by its own losses and thereafter could exert only indirect pressure. He then checked the situation of his own left flank platoon which was supposed to crowd the enemy line of escape leading from the rear of the Strong Point. That platoon had closed in as far as possible; it was under heavy fire and was engaging with its own automatic weapons and light mortars.

It struck Rideout that the deadlock could be broken only by armor. But during the ten days of stalemate the fields surrounding the Kergonant battery had been greatly torn by shellfire; he knew as he surveyed the ground that the tanks could not move forward far enough to rally his own men and help destroy the remaining enemy fractions. So he went after an engineer detail, guided it forward, and showed it where to build a road across the field leading to his forward lines. He ordered his own men to keep the enemy occupied with fire while the engineers worked. The construction job took five hours.

That accomplished, four medium tanks moved up to within approximately 150 yards of the enemy rampart. They were nosing into the defensive wire when two of them were struck by anti-tank shells from within the fortress. The crews were incinerated. This disaster ended the tank action, though it did not defeat Rideout.

One of his squads had been working with the Company to his right, the peculiar nature of the hedgerow country having necessitated this use of the squad as a contact element. Rideout brought this squad back to his own center, then asked for the loan of a platoon from the other Company. It was a move that he might have attempted earlier in the day but for the hope that the right flank company would be able finally to get in motion and cut off the enemy line of retreat.

Now as he made it, it did not add greatly to his own strength. By this time both Companies had been reduced through battle losses to less than 80 men. His slender re-enforcement—the squad and the platoon together—comprised but 23 men. He personally led them forward to the vital point. They arrived crawling, yet even this manner of arrival was enough to restore motion to the stalled line. When they moved through, about three other squads of men joined them. The crawl continued. They got to the shattered walls of the first gun emplacement within the fort. Then they arose with a yell and ran forward and the remaining few of the enemy who were not killed ran toward the sunken road which was the only exit from the works.

Said Rideout: "It was the best day of battle I ever had. It was the only day on which I was in contact with my flanks and rear at all times."

There was the real source of his personal confidence, and in view of all that went wrong elsewhere, can it be doubted that this was the thing essentially which enabled Rideout to score a considerable victory? The fall of Kergonant echoed across the front of a whole army corps. With this obstacle out of the way, the Division was able to advance 11 kilometers during that night and the next day.

The words that he said as he stood in front of his Company telling about the charge were equally to the point. I wrote them down as he said them. I well recall his youthful earnestness; he was but twenty-one at that time.

> In any such desperate action as a charge, it is necessary to have an officer boldly leading. But it is not less important to have one forceful individual remain behind to do the pushing. Otherwise, many of the men will not go. The weak ones always know that they will be able to offer some kind of excuse for their own failure.
>
> I gave myself the task of remaining behind and prodding them because I am the commander. It was my plan and I had given the order. It was my duty to see that it was carried out. I considered that my post should be at the point which offered the best chance of bringing off a successful and completed action. My men know me well enough to have confidence that when I remain behind, it is for the good of the Company and not to save myself.

The attitude of his men supported every word that he said. Moreover, I can add my personal witness that in the course of the past war, I saw many a commander senior to Rideout fail finally of his objective in leading his men in the same kind of action because, while setting a brave personal example, he failed to "close the circuit" in making sure that his troops would comply with the order.

Several weeks after this time, Rideout was killed in action; I believe it was somewhere about Aachen. His span of combat service had been relatively brief. Yet in that time he had acquired the kind of wisdom which counts most in the scales of war. All that he said and did bespoke a keen interest in men. He made it the foremost part of his job to see his own men as they were, both in their strength and in their weakness, seeking to cultivate the one while at the same time remaining clear-eyed to the need for safeguarding against the other. He did not expect too much of them but he demanded the maximum that he thought it was possible for them to give. Then of himself he asked more.

If there is much more to moral leading than is epitomized in this one example set by a novice who had scarcely attained man's stature, then I must admit that in my privileged rounds with combat troops it has escaped me.

Once again, however, it might be well to speak of the importance of enthusiasm, kindness, courtesy, and justice, which are the safeguards of honor and the tokens of mutual respect between man and man. This last there must be if men are to go forward together, prosper in one another's company, find strength in the bonds of mutual service, and experience a common felicity in the relationship between the leader and the led. Loyalty is the big thing, the greatest battle asset of all. But no man ever wins the loyalty of troops by preaching loyalty. It is given him by them as he proves his possession of the other virtues. The doctrine of a blind loyalty to leadership is a selfish and futile military dogma except in so far as it is ennobled by a higher loyalty in all ranks to truth and decency.

War is much too brutal a business to have room for brutal leading; in the end, its only effect can be to corrode the character of men, and when character is lost, all is lost. The bully and the sadist serve only to further encumber an army; their subordinates must waste precious time clearing away the wreckage that they make. The good company has no place for the officer who would rather be right than be loved, for the time will quickly come when he walks alone, and in battle no man may succeed in solitude.

I have known a few brutes in battle whose talents were so limited that they could try no other means of command than the abuse of men. But I have yet to see one who did a good job of holding his command together when the going became rough, and in the ranks fear of the enemy began to eclipse fear of the man up top.

Ruggedness? Toughness? Ah, these are quite different things. So long as they are only the outer reflection of an inner determination and so long as the inner fire is tempered by a heart having real compassion for men, these are the best hands for the business. They will win the hearts of other men and will stimulate their valor. These others will try to be like them, for it is a truth not to be denied that the rugged way is the natural way in battle.

There comes to mind one last picture from the same campaign in which Rideout took Kergonant.

The scene is stone-walled Fort Montbarey, the last obstacle barring entry into Brest. A battalion of the 116th Infantry Regiment under Major Tom Dallas has had the Fort invested for three days.

But the defenders have withdrawn to within the inner walls and will not surrender. The infantry fire cannot get at them. The ports have been flamed but without visible effect. Finally, Dallas asks for three tons of TNT to blow the walls. He is given one ton. Assisted by the infantry, the engineers lay the charge in a gallery under the wall. The work completed, Dallas is ready to give the order. Then he remembers.

On the night before, the Battalion had attacked and Lieutenant Durwood C. Settles had been killed in the moat. There has been no chance to recover the body. It is still there and will be crushed by the falling masonry.

So the demolition is held up and Dallas asks for a volunteer to go down into the moat, under fire, and bring out the body. A young lieutenant named Kelton responds.

In a few minutes, Kelton returns with his burden.

He says to Dallas: "We're ready now."

Dallas replies: "Ready? Then blow them all to hell."

The charge goes off. The earth shakes. The walls collapse with a roar. There is a stunned silence from within the Fort.

Dallas stands there for a few seconds and the tears fall as he looks down at the body of his dead officer.

13

FOOTNOTE TO HISTORY

"I returned and saw under the sun that the race is not to the swift nor the battle to the strong."

—ECCLESIASTES.

IN THE historical operation within the Army of the United States which was begun just before midway of World War II, some mistakes were made and many opportunities were missed because we were all feeling our way after a late start and because we continued to operate without adequate trained man power.

But I think that from the beginning we did one thing soundly. Even before we had outlined a program or surveyed the problem, we saw it as a working principle that a great part of our effort would have to be directed toward recapturing the detailed story of the fighting line. In the hour when we made that decision we had fewer men than can be counted on one hand and the assignment which had been handed us was as vast as the war itself. In these circumstances it would have been following the path of least resistance to say that our whole time would have to be given to the search of operations at the higher levels. There was ample precedent for such a course. The body of military history is almost exclusively a record of the movements of armies and corps, of decisions by generals and commanders-in-chief, of the contest between opposing strategies and the triumph of one set of logistical conditions over another. The occasional rare passages from the battlefront which are thrown in to illuminate and make zestful the story of the over-all struggle are usually of such glittering character or dubious origin to warrant a suspicion that they have little real kinship with the event.

But as we saw our task then, though we did not state it to ourselves in precisely these terms, there was a compelling reason why we should risk missing something at the higher levels to gain more thorough knowledge than has been had before of the acts and nature of our men in combat. War is always an equation of men and machines. Efficiency comes of a proper balancing of the equation.

Because of the great wealth and productive power of our nation, we can afford a system of war which is based on the conserving of men. It is an integral feature of our system that we will spend matériel at any point and in any amount needed to preserve the lives of soldiers. We do not believe in wasting infantry on missions which can better be done by tanks; we are opposed to wasting armored force lives on tasks which can be accomplished by artillery and air bombardment. Though lives are sometimes tossed away in our battles, it is not the American policy to sacrifice men in order to save machines. Even so, the deaths of a quarter million men and the wounding of thrice that number are a reminder that there are limits to the uses of the machine in war and that its efficiency as a saver of human life is according to the efficiency, intelligence, and courage of the relatively few men who must take the final risks of battle.

Though we could not foresee all of the developments which would make the last year of the war seem almost a trial run for the mighty engines of the future, it was clear to us that if there was to be a better balancing of American arms in time to come, there was need for a closer regard of the conduct and nature of the individual man in battle than there had ever been. Even so, this was not a very high target at which to aim our shot. In the United States Army in wars past no means had existed for collecting the facts and phenomena of the battle line. Post-combat investigation of cause and effect in the small picture of war was no part of the duties of the General Staff.

The idea of working out the processes of the future from the occurrences of the past was by no means foreign to the professional concept of the art of war, but it was never assiduously applied to operations at the lower levels and the search had not been systemmatized in the higher levels. Yet primarily it was not the soldiers who were at fault. In time of peace the Army of the United States was never so generously used by its people and government that it could prepare for such enlightened enterprises when war came, and after the rise of any great emergency, the attention of the General Staff had to be directed toward more vital matters. In consequence, the battle experience of Americans has always been left the property of private individuals rather than treated as a public asset, to be applied for the benefit of the Army as a whole.

We believed that this situation should be corrected as far as lay within our power, though we knew that toward that end we would have to direct a greater portion of our time than might otherwise be desirable to the first-hand study and evaluation of operations along the line of fire. This search was nominally the responsibility of no other General Staff section. It was thus almost by default that our mission came to include a type of research which is so closely linked to both the field operations and training aspects of the G3 function that one cannot but feel that the latter is incomplete without it. In what we did we had always the active support of this and other General Staff sections. It was not lack of imagination which had held them from working along parallel lines but lack of personnel. We too were under this same pressure. There was never the slightest chance that our slender resources of man power could be stretched to cover all battlefronts. Some actions we had to pass entirely in order to become reasonably complete with others. We worked always in the hope that we could do the requisite number of studies sufficiently well to demonstrate the values of a continuing tactical research during battle, thereby commending the process to the thoughtful regard of the General Staff.

These studies have not been made a part of this book. They nonetheless form the basis, along with my prior experiences in service and my lifelong interest in the military problem, of all that has been said here. I have drawn only from my own written materials as these are the episodes that I know best. It is unavoidable that experiences which one has shared with living men come to have far clearer meaning than the most forceful impression which has been written by another. As I look once again over my old notes, I see once again the faces of many of these men, and I recall many of their words and the earnestness with which they were uttered. There would be no point in trying to dissociate one's self from such memories thereby to deal coldly and mathematically with the notes as if they were the work of some other hand. The validity of an idea comes finally of one's grasp of all of the attendant circumstances. Out of these actions I know which ones tried men to the soul because of the fierceness of the fire which came against them and I know equally well the other ones wherein the danger was more apparent than real. Such was the pressure of our work that one rarely had time to record the whole body of impressions which bore upon one's understanding of the event. But they became nevertheless an indispensable part of the process of assimilating the truths of battle, of weighing its phenomena, and of distinguishing between those things which were special to one situation and those which were general to all combat. It is my belief that there is no better school for this than the experience of dealing first-hand with the participants in a thousand or more skirmishes.

It was partly by the accident of assignment and partly through choice that a greater share of this work became my portion than fell to any other individual and that having pioneered it, I was able to continue until the close of the war. This entailed an expenditure of time which might better have been directed to higher administrative duties, but as these latter always seemed the better for my absence, I doubt that the service was any the loser. In the course of my official duty, however, I came to know the historical operations in the British, French, and Canadian Armies, and after the fighting ceased, several historical officers of the German Army and a large number of its higher commanders came temporarily into my custody. Thus I speak not out of ignorance in saying that while we may have been laggard in the Army of the United States in developing an historical process which embraced the battle line and contended directly against its fog, we have still done better than any other army in this particular.

In my collected thinking about my experiences with battle troops there is one lasting impression which stands above all others. As a student of military history my reading between wars had made me overrespectful of the factor of the preponderance of force in warfare. I came to believe that battles and campaigns were almost invariably won according to which side was in position to apply the greatest weight at the decisive point. That is perhaps a relative truth. But once one falls in love with this idea, it is only a short step to a wholly materialistic concept of the balancing of power and the making of military decision. Success becomes a purely mathematical problem of counting men and machines and what is required to supply them. I know now that that is not true. If I learned nothing else from the war, it taught me the falseness of the belief that wealth, material resources, and industrial genius are the real sources of a nation's military power. These things are but the stage setting; those who manage them are but the stage crew.

The play's the thing. Finally, every action large or small is decided by what happens up there on the line where men take the final chance of life or death. Though I would not for a moment contend that modern war can be fought and won without vigorous thought and action on the home front, I deny absolutely that these things can vouchsafe military victory any more now than in the days when men fought with spears and crossbows. Any who look at war and think otherwise are sighting through the wrong end of the telescope. They have become deceived by the vastness of the national preparation. How differently they would see things if it became their duty to measure the thin margins between victory and defeat on the field itself!

As I went about my work, I came to see, more fully and more surely than I have expressed it in the tactical portions of this book, that the great victories of the United States have pivoted on the acts of courage and intelligence of a very few individuals. The time always comes in battle when the decisions of statesmen and of generals can no longer effect the issue and when it is not within the power of our national wealth to change the balance decisively. Victory is never achieved prior to that point; it can be won only after the battle has been delivered into the hands of men who move in imminent danger of death. I think that we in the United

States need to consider well that point, for we have made a habit of believing that national security lies at the end of a production line. Being from Detroit, I am accustomed to hearing it said publicly that Detroit industry won the war. This may be an excusable conceit, though I have yet to see a Sherman tank or Browning gun that added anything to the national defense until it came into the hands of men who willingly risked their own lives. Further than that, I have too often seen the tide of battle turn around the high action of a few unhelped men to believe that the final problem of the battlefield can ever be solved by the machine. We are all looking for some assurance for the national future. We want to know that there is a main chance for continuing the well-being of our kind of society.

But where does it lie? If the people are right who put the accent on America's great industrial power and inventive genius, then I am all wrong about it, and I am convinced that the stark facts of the battlefield are wrong likewise. What I learned there taught me once again that courage is the real driving force in human affairs and that every worth-while action comes of some man daring what others fear to attempt. Freedom rests on a base composed of men and women who can agree with Marcus Aurelius that "It is a useful help toward contempt of death to pass in review those who have tenaciously stuck to life." But I do not believe this lesson is to be learned by a people which calculates its strength by counting its machines or that a society which comes finally to stake its hope only in such tabulations can keep alive the vital spark which is necessary to its security.

So it comes down to this—that the ideal of automatic response is as impractical for the American nation as for its foot soldiers, those sons of toil who we in our blindness speak of as "the poor doughboys," though they have the right to be the proudest men who walk our soil.

While my primary object in this book has been to remark on those matters which are vital to the efficiency of men in combat, there is a point at which they become inseparable from the deep currents which should give meaning to the national life.

The man who is willing to fight for his country is finally the full custodian of its security. If there were no willing men, no power in government could ever rally the masses of the unwilling. But if the spirit and purpose which enable such men to find themselves and to act are to be safeguarded into the future, much more will have to be required of the country than that it point its young people toward the virtues of the production line. There is something almost fatally quixotic about a nation which professes lofty ideals in its international undertakings and yet disdains to talk patriotism to its citizens, as if this were beneath its dignity.

It was not beneath the dignity of Brigadier General Anthony C. McAuliffe to speak to his men in words of purest patriotism during the darkest hours of the siege of Bastogne. That line from his famous message to his troops: "We are giving our country and our loved ones a Christmas present," has the chime of a silver bell. And he was speaking to a particularly hardboiled group of Americans, engaged to the hilt in a fight for survival.

Perhaps it is at least a part of our failure that we love to strike the grand pose. We like to pretend an intellectual maturity which refutes the passions that have shaped the destinies of other nations. We feel that this makes us better than nations which believe in national spirit, and that in some unexplained fashion it also makes us less vulnerable. A note of smugness was not missing from the remark all too frequently heard during World War II: "We go at this thing just like it was a great engineering job." What was usually overlooked was that to the men who were present at the pay-off, it wasn't an engineering job, and had they gone about their duty in that spirit, there would have been no victory for our side.

To men who have been long in battle and have thought about it deeply, there comes at last the awareness of this ultimate responsibility—that one man must go ahead so that a nation may live. No feeling of futility accompanies that thought. At the time one accepts it simply as the rule of life and of death, of struggle and of national survival. But in the long afterglow comes also the realization that a nation may perish because too few of its people have found this truth on the only field where it may be found and because too many of the others who have not found it, unconsciously resist it and rule it out because it has not been part of their experience.

And so the final and greatest reality, that national strength lies only in the hearts and spirits of men. The Army, Navy, and Air Force are not the guardians of the national security. The tremendous problem of the future is beyond their capacity to solve. The search begins at the cradle where the mother makes the decision, either to tie her child to her apron strings or to rear him as a man. It continues through the years of schooling when children are taught either to place personal interests uppermost or to think in terms of their responsibility toward their society, their country, and all of mankind. It carries into the halls of government where our lawmakers may vote either to awaken our youth to a new understanding of duty or to continue the indulgent course which is more likely to find favor with the majority of their constituents.

INDEX